Tourism and Transition

Governance, Transformation and Development

DEDICATION

To Lily and Bob
1913–2002/2001

Tourism and Transition

Governance, Transformation and Development

_____

Edited by

Derek R. Hall

Professor of Regional Development
The Scottish Agricultural College (SAC)
Auchincruive, Ayr KA6 5HW, UK

CABI Publishing

CABI Publishing is a division of CAB International

CABI Publishing
CAB International
Wallingford
Oxfordshire OX10 8DE
UK

Tel: +44 (0)1491 832111
Fax: +44 (0)1491 833508
E-mail: cabi@cabi.org
Website: www.cabi-publishing.org

CABI Publishing
875 Massachusetts Avenue
7th Floor
Cambridge, MA 02139
USA

Tel: +1 617 395 4056
Fax: +1 617 354 6875
E-mail: cabi-nao@cabi.org

A catalogue record for this book is available from the British Library, London, UK.

Library of Congress Cataloging-in-Publication Data
Tourism and transition: governance, transformation, and development/edited by Derek R. Hall
        p. cm.
Includes bibliographical references and index.
    ISBN 0-85199-748-1 (alk. paper)
    1. Tourism. 2. Tourism--Social aspects. 3. Economic development.
I. Hall, Derek R.
    G155.A1T589525 2003
    338.4'791--dc22

                            2003020954

ISBN 0 85199 748 1

Typeset in 9pt Souvenir by Columns Design Ltd, Reading
Printed and bound in the UK by Cromwell Press, Trowbridge

Contents

List of figures

Photographic contributions are by the chapter's first-named author unless otherwise indicated in
the above list.

List of tables

Contributors

Habib Alipour, School of Tourism and Hospitality Management, Eastern Mediterranean University, Famagusta, (Turkish Republic of North) Cyprus

Maria Attard, Geography Division, Mediterranean Institute, University of Malta, Msida, Malta

Zsuzsanna Behringer, Directorate for Marketing Research and Analysis, Hungarian National Tourism Office, Budapest, Hungary

Jenny Briedenhann, New Economics Foundation, London, UK

Steve Butts, Faculty of Land, Food and Leisure, Seale-Hayne, University of Plymouth, Newton Abbot, UK

Derek Hall, Tourism Research Group, Research and Development Division, The Scottish Agricultural College, Auchincruive, UK

Rong Huang, Centre for International Tourism, Hospitality and Leisure Research, University of Derby, Derby, UK

Hasan Kilic, School of Tourism and Hospitality Management, Eastern Mediterranean University, Famagusta, (Turkish Republic of North) Cyprus

Kornélia Kiss, Directorate for Marketing Research and Analysis, Hungarian National Tourism Office, Budapest, Hungary

Wacław Kotlinski, Faculty of Economics, University of Technology, Rzeszow, Poland

Jovan Popesku, Centre for Responsible and Sustainable Tourism Development, College of Tourism, Belgrade, Serbia

Lesley Roberts, Centre for Travel and Tourism, Business School, Northumbria University, Newcastle-upon-Tyne, UK

Peter Schofield, Marketing and Tourism Management, School of Leisure, Hospitality and Food Management, University of Salford, UK

Antonis Theocharous, Iannis Gregoriou Business School, The Cyprus College, and Research Laboratory for Crisis and Risk Management, Nicosia, (Republic of) Cyprus

Barry Worthington, Tourism Research Fellow, Abertay University, Dundee, UK

Takayoshi Yamamura, Department of Tourism Design, Kyoto-Saga University of Arts, Kyoto, Japan

Preface

This volume has its origins in the international tourism conference panel 'Tourism, transition and global capital' hosted by the University of Wales Institute, Cardiff (UWIC) in September 2002. The editor had the pleasure of chairing the panel, which, while the smallest of the conference in terms of paper contributions, was one of the more lively groups and revealed an internal coherence which immediately suggested itself in the form of a closely edited and integrated text.

The panel was refreshing both for the variety of new faces and voices that it was able to bring to the debate, and for the wealth and diversity of empirical interest that the papers articulated.

Although not always easy to edit, the collection that has evolved into this volume represents original thinking from a number of new directions for an English language audience.

At a personal level the conference provided me with the first opportunity to get back into the public arena following the death of my mother, who, after 64 years of devoted married life, had been unable to cope with the death of my father less than 9 months previously. This volume is dedicated to their memory.

I have particularly enjoyed working with the contributors in bringing this volume to fruition and in learning a little more about them. Thanks are due to them all, especially to those contributors who completed their manuscripts within a relatively tight schedule, and to those who did so under not always easy circumstances. The conference organizing committee at UWIC and especially Claire Haven deserve thanks for setting the circumstances from which this book could develop, and for inviting me to chair the panel. Appreciation is also extended to colleagues at CAB International for their efficient management of the publication of this volume, and in particular to Rebecca Stubbs for her support and encouragement.

I would like to express my special thanks to Irene Kirkpatrick who meticulously word-processed most, and particularly some of the more heavily edited chapters, who patiently formatted the work, and who consistently highlighted the not inconsiderable shortcomings in first draft editing. And finally to Frances Brown, for her professional support and continued tolerance on matters domestic.

Derek Hall
Auchincruive and Maidens, Ayrshire

Abbreviations and acronyms

$	dollar
acquis communautaire	the common law of the EU
AIEST	International Association of Scientific Experts in Tourism
asl	above sea level
avis	EU opinion (on accession applications)
bn	billion
CAAC	Civil Aviation Administration of China
CDF	comprehensive development framework
CEC	Commission of the European Communities
CEE	Central and Eastern Europe
CEECs	Central and Eastern European countries
CIA	Central Intelligence Agency (USA)
CIS	Commonwealth of Independent States (the states of the former Soviet Union, minus Estonia, Latvia and Lithuania)
CITS	China International Travel Services
CNTA	China National Tourism Administration
CPC	Communist Party of China
CPPS	commuted parking payment scheme (Republic of Malta)
CPI	consumer price index
CSO	Central Statistical Office (Republics of Hungary and Poland)
CTO	Cyprus Tourism Organisation
CTS	China Travel Services
CYTS	China Youth Travel Services
DEAT	Department of Environmental Affairs and Tourism (Republic of South Africa)
DG	Directorate General (of the European Commission)
€	euro
EBRD	European Bank for Reconstruction and Development
EC	European Commission/Community
ECB	European Central Bank
ECU	European Currency Unit
EEC	European Economic Community
EHS	Estonian Heritage Society
EIB	European Investment Bank
EIU	Economist Intelligence Unit

ERDF	European Regional Development Fund
ESF	European Social Fund
ETC	European Travel Commission; English Tourism Council
EU	European Union
EU15	The 15 EU members prior to the fifth (2004) enlargement (Austria, Belgium, Denmark, Finland, France, Germany, Greece, Ireland, Italy, Luxembourg, the Netherlands, Portugal, Spain, Sweden and the UK)
FAO	(United Nations) Food and Agriculture Organization
FDI	foreign direct investment
FSU	former Soviet Union
G-24	World's 24 richest aid donors: Canada, the USA, Japan, Australia, New Zealand, Austria, Belgium, Denmark, Finland, France, Germany, Greece, Iceland, Ireland, Italy, Luxembourg, the Netherlands, Norway, Portugal, Spain, Sweden, Switzerland, Turkey and the United Kingdom
GDP	gross domestic product
GNP	gross national product
ha	hectare
HNTA	Hunan Tourism Administration
HUF	Hungarian *forint* (currency)
IBRD	International Bank for Reconstruction and Development (World Bank)
ICG	institutions, competition and government
ICOMOS	International Council on Monuments and Sites
IMF	International Monetary Fund
JISC	Joint Information Systems Committee (support for UK further and higher education)
KITOB	Cyprus Turkish Hoteliers Union
KITSAB	Cyprus Turkish Travel Agents Union
km	kilometre
KSATS	Kyrgyz State Agency of Tourism and Sport
LDCs	less developed countries
m	metre
MAB	UNESCO Man and the Biosphere Programme (see <http://unesco.org/mab>)
MAPE	mean absolute percentage error
MEPA	Malta Environment and Planning Authority
METCO	Malta External Trade Corporation
MFA	Ministry of Foreign Affairs (Malta)
MIMAS	Manchester Information and Associated Services (UK research and national data centre)
mn	million
MNC	multinational company
MSDPM	Ministry of State and Deputy Prime Ministry (Turkish Republic of North Cyprus)
MTA	Malta Tourism Authority
MTE	Ministry of Tourism and Environment (Turkish Republic of North Cyprus)
NATO	North Atlantic Treaty Organization
NIC	newly industrializing/industrialized country
NGO	non-governmental organization
NSCKR	National Statistical Committee of the Kyrgyz Republic
NSO	National Statistics Office (Malta)

NTC	National Tourism Commission (People's Republic of China)
OECD	Organization for Economic Co-operation and Development
PEV	ProEuropa Viadrina (German–Polish Euroregion)
PHARE	Poland/Hungary Assistance for Economic Reconstruction (EU programme extended to much of the rest of CEE including the Baltic states)
PPP	purchasing power parity
POLINST	political instability database (compiled by Antonis Theocharous)
PRA	participatory rural appraisal
Ramsar	Intergovernmental Convention on Wetlands, signed in Ramsar, Iran, in 1971 (see <http://www.ramsar.org/>)
RDC	regional district council (Republic of South Africa)
RMB	renminbi (collective term for Chinese currency, yuan)
SAC	Scottish Agricultural College
SAS	software manufacturer brand (Holland Numerics)
SEA	South-east Asia
SEE	south-eastern Europe
SGATT	State General Administration of Travel and Tourism (People's Republic of China)
SMEs	small- and medium-sized enterprises
SPA	State Property Agency (Republic of Hungary)
SPO	State Planning Organization (Turkish Republic of North Cyprus)
sq	square
SLP	stabilization, liberalization and privatization
TACIS	EU programme equivalent to PHARE (qv) for the states of the CIS (qv)
TEMPUS	EU higher education exchange and development programme for CEE
TRC	Truth and Reconciliation Commission (Republic of South Africa)
TRNC	Turkish Republic of North Cyprus
TTI	Travel and Tourism Intelligence
UK	United Kingdom (of Great Britain and Northern Ireland)
UN	United Nations
UNDP	United Nations Development Programme
UNESCO	United Nations Education and Science Organization
UNICEF	United Nations Children's Fund
US(A)	United States (of America)
USAID	United States Aid Agency
VAT	value-added tax
VRT	vehicle roadworthiness test
VRTT	virtual reality tourism and travel
WHS	World Heritage Site
WHO	World Health Organization
WTO	World Tourism Organization, World Trade Organization
WTTC	World Travel and Tourism Council

1 Introduction

Derek Hall

The starting point and basic organizing framework for this volume is the oft understated, yet obvious truism, that international tourism is part of a much wider set of social, cultural, economic, political and environmental change agents, and should be viewed within this wider context and not in isolation. As a corollary, it is consequently often difficult to distinguish the impacts of tourism from the influences of other dynamic processes, for such purposes as informing policy. The change agents within which tourism is often embedded are largely driven by the global expansion and rejuvenation of capitalism, a process that inherently generates patterns of uneven development and inequality, which tourism may act to ameliorate or exacerbate. This may take place, for example, through the integration of destinations into dynamic domestic and global markets and networks of international capital. This interface has tended to result in a reinforcement of the economic dominance of metropolitan regions, and differentiation amongst poorer regions, such as urban-industrial rust belts, peripheral and certain rural areas. Further, spatial and structural distortions in transforming economies have tended to focus activity – not least tourism – on major urban areas in favoured regions. Urban tourism in some major cities such as Prague and Budapest has been transformed from a niche to a mass activity, exacerbating congestion and infrastructural problems as well as distorting

territorial development (e.g. Simpson, 1999). Such spatial distortion is likely to intensify within new world (dis)orders: for example, as Central and Eastern European (CEE) cities both within and outside the EU attempt to redefine themselves as major European cultural destinations (e.g. Blonski, 1998).

What is particularly salient for this volume is the ways in which tourism intersects with contemporary processes of capitalist expansion as

Fig. 1.1. Czech Republic: postcard sales in Prague.

Fig. 1.2. Hungary: the national parliament building in Pest viewed from the walls of Old Buda across the Danube.

expressed within and viewed through societies 'in transition'. The most obvious contexts for these interrelationships are:

- the adoption of ('Western') democratic political processes and market economies in some, but not all, former Soviet bloc societies;
- European Union (EU) enlargement to incorporate some European post-communist societies, while whetting the appetite and aspirations of others; and
- the adoption of market-related economic systems within persisting one-party communist political systems.

Additionally, adaptive forms of capitalism, responding to the governance of particular local, regional or national social and cultural characteristics, can be viewed through the lens of the governance of tourism development in:

- post-apartheid South Africa;
- post-division Cyprus;
- Malta, where issues of EU accession represent a substantial political cleavage within society; and

- the understanding and response to impacts of political instability and terrorism.

These are the environments addressed in this volume. This chapter, in combination with the next, aims to establish the frameworks within which their examination takes place. As a first step, it appeared appropriate for an edited text on tourism and transition to open by postulating a model for tourism's own transition, and to offer a view on the likely condition of post-transition global tourism. Second, the chapter addresses the nature and role of economic and political transition within the context of evolving global capitalism and tourism's role within it. Third, it briefly introduces each of the case study chapters (3–14) that follow.

Tourism's own Transition

In the context of this book's focus on processes of tourism development in 'transitional' societies, there is perhaps an initial requirement to address tourism's own transition: to attempt to view generically tourism change and development within a societal frame. Notwithstanding the value of resort destination life-cycle models, the way in which political and/or ideological influences interact with tourism development may often only be evident at a societal level. But societal level analysis has perhaps been neglected compared with that for destination resorts.

In attempting to offer a simple conceptual model of the tourism transition – the process through which tourism development in any given society passes to reach a common state of dynamic equilibrium – it is posited that there are two central multi-dimensional sets of characteristics influencing the path of tourism development, as set out in Fig. 1.3:

- the societal context within which tourism takes place, in terms of the general level of economic development of, and dominant ideological impress on society; and
- the structural and spatial nature of the tourism industry itself, in terms of dominant mode of tourism activity, tourism industry organizational structure, scale and spatial characteristics of tourism activity, and relationships between tourism product supply and demand.

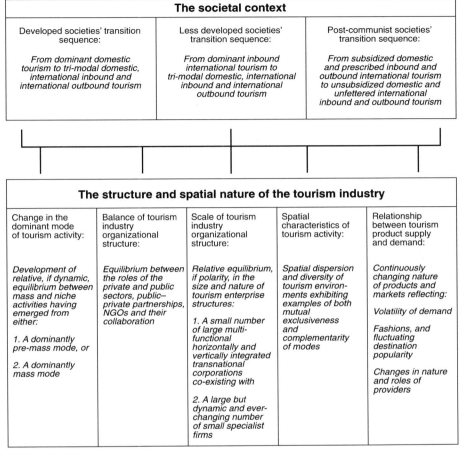

The societal context		
Developed societies' transition sequence: *From dominant domestic tourism to tri-modal domestic, international inbound and international outbound tourism*	Less developed societies' transition sequence: *From dominant inbound international tourism to tri-modal domestic, international inbound and international outbound tourism*	Post-communist societies' transition sequence: *From subsidized domestic and prescribed inbound and outbound international tourism to unsubsidized domestic and unfettered international inbound and outbound tourism*

The structure and spatial nature of the tourism industry				
Change in the dominant mode of tourism activity:	Balance of tourism industry organizational structure:	Scale of tourism industry organizational structure:	Spatial characteristics of tourism activity:	Relationship between tourism product supply and demand:
Development of relative, if dynamic, equilibrium between mass and niche activities having emerged from either: *1. A dominantly pre-mass mode, or* *2. A dominantly mass mode*	*Equilibrium between the roles of the private and public sectors, public–private partnerships, NGOs and their collaboration*	*Relative equilibrium, if polarity, in the size and nature of tourism enterprise structures:* *1. A small number of large multifunctional horizontally and vertically integrated transnational corporations co-existing with* *2. A large but dynamic and everchanging number of small specialist firms*	*Spatial dispersion and diversity of tourism environments exhibiting examples of both mutual exclusiveness and complementarity of modes*	*Continuously changing nature of products and markets reflecting:* *Volatility of demand* *Fashions, and fluctuating destination popularity* *Changes in nature and roles of providers*

Note: this model was developed jointly with Lesley Roberts; her contribution and support are gratefully acknowledged.

Fig. 1.3. The tourism transition model.

The two essential components of tourism's transformation which temper the 'end' state of its transition are:

- equilibrium: the extent and nature of the structural and spatial balance of and between tourism development and its context; and
- dynamism: a process of continuous change whereby the equilibrium of these relationships is far from static but reflects change, tension and even conflict – over, for example, tourism development aims and objectives, or resource access issues – within and between the tourism industry and its wider societal context.

In this way tourism is sustained within the changing circumstances of any given society, although this may say little about tourism's wider sustainability role within that society's development processes.

The essential characteristic of the tourism transition is thus transformation from structural imbalance and spatial distortion to a dynamic equilibrium within which there is:

- a structural and spatial balance of mass and niche activities, products and infrastructures; and
- a convergence of domestic and international modes expressing such structural and spatial balance.

Tourism's post-transition environment, which may be conceived of as approximating to Urry's (1995) 'post-tourism' condition, raises serious global questions. During the 21st century geopolitical issues of access to travel and recreation resources are likely to become an important focus of potential tension and conflict in the face of heightening global pressures. The second half of the last century witnessed the growing significance of conflict and potential conflict over access to resources which took on a 'North–South' dimension, and was characterized by the developed world lecturing the governments of less developed countries (LDCs) that they should:

- control their population growth in order to conserve resources, even though the developed world passed through its high population growth phases to its demographic transition largely unrestrained;
- conserve their natural resources, such as tropical rainforests, and reduce pollution and emission of greenhouse gases for the good of the planet, even though the developed world continues to consume a far greater level of resources *per capita* than developing countries (and in a number of instances continues to encourage, both explicitly and covertly, the unsustainable exploitation of LDC natural resources for its own consumption, including those of tourism); and
- restrain migration and refugee movements – a contemporary issue with east–west as well as south–north dimensions, long foreshadowed both in academic analysis and popular fiction (and in some instances influenced by tourism development).

In a parallel expression of global inequality, until now most international tourism has been undertaken by the residents of the economically more developed countries. However, domestic tourism and recreation in less developed countries such as China and India is numerically substantial, and on both a spatial and numerical scale at least as great as much of the international tourism experienced in Europe: witness the 30 million or more attendees from across India at the 42-day-long 2001 Kumbh Mela Hindu festival on the Ganges at Allahabad.

Although at a global level there remain obvious constraints on certain types of activities and restrictions on access in certain areas, generally there is relative freedom of movement and access around the globe, subject only to the ability to pay. But such freedoms are available largely because they are only able to be enjoyed by a restricted proportion of the world's inhabitants. What will be the result when, later this century, living standards have risen sufficiently for the one and a half billion Chinese, one billion Indians and hundreds of millions of Indonesians, Brazilians and other residents of currently 'less developed' countries to pass through the tourism transition and become global tourists?

We are told that skies are dangerously overcrowded, increasing numbers of aircraft are contributing to ozone layer depletion, and that many urban and rural attractions are diminished by congestion, many irrevocably degraded. Will tourism authorities see their roles reversed: rather than encouraging, actually dissuading (at least certain types of) tourists from visiting? Of course, to some extent this has been taking place for some time: many tourism authorities and enterprises target markets for 'quality' rather than quantity – smaller numbers of high-spending, relatively short-stay tourists being favoured rather than larger numbers of low-spending longer-stay ones. Further, some destination areas – certain alpine valleys – and countries, such as Bhutan, restrict absolute visitor numbers based on arguments of carrying capacity and sustainability.

However, such approaches are relatively fragile current palliatives, raising the question of the likelihood of available substantive valves for relieving the pressures of post-transition mega-mass global tourism. Three avenues might suggest themselves.

- Virtual reality travel and tourism (VRTT): although a great deal depends on perfecting the technology (e.g. Buchroithner, 2002), particularly the ability for two or more people to share virtual experiences (e.g. Liebman Parrinello, 2001), research and conventional wisdom suggests, on balance, that the experience of VRTT and the use of real-time webcams merely whets appetites to undertake actual travel and tourism rather than offering a serious alternative to it (Cheong, 1995; Williams and Hobson, 1995).

- Sub-marine travel and tourism: although two-thirds of the earth's surface is water, it is mostly unexplored due to the limits on our ability to develop technology sufficient to withstand the physical pressure – both on material and human bodies – at depth. Can future improvements in technology help us to reduce competing pressures on scarce land resources and divert recreation and even food production to sub-marine environments (e.g. Glowka, 2003)? And if so, will familiar conflicts over administrative jurisdiction arise when resources become exploitable? Or will the ocean simply act as an escape for those few who can afford to get away from land-based mass recreation and the excesses of climate change? Or, indeed, will the perturbations of global warming render sub-marine endeavours more difficult than they are at present?

- Space travel and tourism: the precedents of two individual space tourists and the fact that a number of companies have accepted reservations for sub-orbital trips might suggest the potential for large numbers of people to travel outside the earth's atmosphere. However, even if issues of disposing of human waste products and questions of space sickness can be overcome in an acceptable manner, the cost of technology – particularly re-usable launch vehicles – the ethics of, and relative hostility of space authorities towards, the use of space for recreational purposes, remain obstacles to anything more than occasional individual space tourism activity by the extremely wealthy (Williamson, 2003; Brown, 2004).

Strictly limited by technology, political will and ability to pay, these potential developments are likely to be accessible to relatively small numbers and will have no significant role as safety valves. Actual terrestrial mega-mass tourism will remain the predominant post-transition global mode. 'Niche' tourism, while specialized, will take on mass proportions – as numbers involved in rural recreation and urban tourism already reveal. Consequently, increasing pressures will demand new strategic and imaginative thinking.

Will the protective blocs of developed countries, to defend their own resource interests, argue again that less developed countries be restricted in their access to international travel and tourism when those from the developed world have been able to travel the globe for decades relatively unfettered? Could there ever be consensus at a global level to ration travel and access? If so, how would such a politically sensitive agreement be achieved? Who would arbitrate? The United Nations? How influential would the World Tourism Organization be in the face of super-state (and mega-corporation) power muscle flexing? Or will rationing continue to take place *de facto* on the basis of ability to pay, with consumer costs rising partly to pay towards the environmental costs incurred in global travel? Would a belated 'polluter pays' approach thereby maintain the in-built dominance of the more developed world with its initial advantage and higher living standards? Or will travel and recreation continue unabated, partly 'democratized' by web-based yield management air fare pricing mechanisms (e.g. Klein and Loebbecke, 2003), until there is 'global gridlock' and conflict over access to international travel and recreational resources becomes inevitable? Or will some unforeseen Malthusian break, even global terrorism, intervene, irrevocably?

Context

With one eye on such considerations, the purpose of this volume is to bring together current research on the inter-relationships between tourism, governance and development within contexts of 'transition'. It embraces studies of tourism development within contemporary 'transitional' pathways in societies that are:

- emerging or have emerged from state socialism,
- experiencing economic and/or political restructuring processes, and
- adjusting to the accession requirements of the EU.

In some cases, two or even all three of these contexts inform the chapters that follow (Table 1.1). Thus for the purposes of this volume,

Table 1.1. Selected countries in transition: key indicators, 2000/2001.

Country	Total population (mn)[f]	Area ('000 sq km)	Population density (per sq km)	GDP PPP ($bn)	GDP per capita PPP ($)	Unem- ployment rate (%)	Life expectancy at birth		Infant mortality rate (per 1000 live births)	Estimated Internet users per 1000 pop.	Tourism receipts as % of GDP
							Female	Male			
CEE post-Communist 2004 EU entrants											
Czech Republic	10.2	78.9	130	154.2	15,064	8.1	78.5	72.1	4.1	98	0.1
Estonia	1.4	45.2	30	13.7	10,049	12.6	76.1	65.2	8.4	272	0.6
Hungary	10.2	93.0	110	129.6	12,728	5.7	75.6	67.1	9.2	145	0.03
Latvia	2.3	64.6	37	18.5	7,809	13.1	76.0	64.9	10.4	62	3.0
Lithuania	3.5	65.3	53	29.1	8,359	17.0	77.9	67.6	8.6	61	0.9
Poland	38.6	312.7	124	380.5	9,844	17.4	78.0	69.7	8.1	72	0.9
Slovakia	5.4	49.0	110	66.2	12,314	19.2	77.2	69.1	8.6	120	0.05
Slovenia	2.0	20.3	98	35.4	17,762	5.9	79.1	71.9	4.9	151	de
Mediterranean 2004 EU entrants (including Northern Cyprus)											
Cyprus	0.8	9.3	82	9.1	11,567	4.0	80.4	75.3	5.6	176	34.4
Malta	0.4	0.3	1,239	3.6	9,255	5.0	80.3	76.0	6.0	130	39.4
Other CEE (including Turkey)											
Albania	3.2	28.8	119	12.8	3,743	14.5	78.0	72.0	11.6	1	0.1
Belarus	9.9	207.6	48	84.7	8,422	2.1	74.5	62.8	9.3	18	<0.01
Bosnia-Hercegovina	4.1	51.2	84	11.7	2,626	39.9	74.7	68.7	15.0	na	0.2
Bulgaria	7.9	110.9	72	50.6	6,366	19.4	75.3	68.2	13.3	52	4.0
Croatia	4.4	56.5	78	36.0	8,118	15.8	76.7	69.1	7.4	56	1.8
Georgia	5.2	69.7	72	15.7	3,553	15.8	77.4	71.9	14.9	4	6.9
Macedonia, FYR	2.0	25.7	79	9.9	4,882	30.5	74.8	70.5	11.8	25	0.4
Moldova	4.3	33.9	108	8.4	2,307	7.3	71.2	63.9	18.4	12	0.3
Romania	22.4	238.4	94	156.3	6,976	7.3	74.2	67.0	18.6	172	<0.01

Russia	144.1	17,075.4	1,229.0	8	8,490	8.9	72.3	59.0	14.6	7	0.1
Serbia/Montenegro	10.7	102.2	18.1[b]	104	3,532	21.2	76.7	70.6	13.0[b]	na	na
Turkey	69.6	774.8	399.1	87	5,901	8.5	72.5	66.8	36.6	30	<0.01
Ukraine	48.7	603.7	204.1	81	4,155	11.1	73.6	62.3	12.0	6	1.2
Other Former Soviet Union											
Armenia	3.1	29.7	10.8	128	2,839	9.8	74.3	70.3	15.8	17	<0.01
Azerbaijan	8.2	86.6	24.2	94	2,982	1.3	75.1	68.1	12.8	2	<0.01
Kazakhstan	14.8	2,724.9	101.9	5	6,871	2.8	71.6	60.2	19.1	6	0.01
Kyrgyzstan	5.0	199.9	13.5	25	2,724	3.2	72.4	64.9	23.0	11	0.02
Tajikistan	6.3	143.1	7.9	44	1,260	2.6	73.8	69.8	17.8	0.5	na
Turkmenistan	5.5	488.1	33.4	11	4,661	na	70.4	63.4	52.0	1	<0.01
Uzbekistan	25.4	447.4	63.2	56	2,532	0.6	72.2	67.0	19.1	5	<0.01
Other Asian											
Cambodia	12.8	181.0	18.7	71	1,500	2.8[e]	59.5[f]	54.8[f]	64.0[f]	2[f]	1.2
China	1,281.0	9,597.0	5,732.0	133	4,600[f]	3.1	73.9[f]	70.0[f]	27.3[f]	46[f]	0.3
Korea, DPR	22.2	120.5	22.0[f]	184	1,000[f]	na	74.4[f]	68.3[f]	22.8[f]	na	na
Lao PDR	5.8	236.8	9.2	25	1,630	5.7[c]	55.9[f]	52.0[f]	91.0[f]	3[f]	1.2
Mongolia	2.7	1,565.0	4.7	2	1,770	20.0	66.9[f]	62.5[f]	52.0[f]	17	0.6[e]
Vietnam	81.1	329.6	168.1	246	2,100	25.0[a]	72.5[f]	67.4[f]	29.3[f]	19 f	0.05[d]
Post-apartheid											
South Africa	43.6	1,219.9	441.6	36	9,400	37.0 [f]	45.7[f]	45.2	61.8[f]	65	0.6[e]
Latin Caribbean											
Cuba	11.3	110.9	25.9 f	102	2,300f	4.1	79.2[f]	74.2[f]	7.3[f]	11	6.8

Key: [a]1995, [b]1996, [c]1997, [d]1998, [e]1999, [f]2002 estimates.
de, apparent data error in source; *na*, data not available.
Sources: CIA, 2003; United Nations, 2003a,b; UNECE, 2003; World Bank, 2003; WTO, 2002a,b; author's additional calculations.

'transition' usually implies change in political and economic and thus social and cultural dimensions of any society. For Gready (2003: 2), for example, 'transition' from a political science perspective, is change 'towards greater democratization, post-repression, post-colonialism, post-war', which is defined and delimited by institutional mechanisms in a particular way. Yet the reality may be:

- a contested and incomplete process,
- characterized by often personalized considerations of power and access to it,
- incorporating both continuity and uneven change, and
- continuing patterns from the past which may simply be reconfigured rather than being radically changed (Gready, 2003: 2).

Further, the process of transition may witness tensions and potential conflict between local, regional, national and global needs and priorities.

Following the demise of communism in Central and Eastern European countries (CEECs), the Soviet Union and Mongolia, 'transition', an economic project with a strong political agenda, became embraced as the mantra for, and was appropriated to become virtually synonymous with, post-communist change in the former Soviet bloc states. Despite the fact that the former USSR was, geographically, a predominantly Asian country, the focus of 'transition' was markedly Eurocentric, with an ideological driving force that was strongly trans-Atlantic, in the wake of the Reagan–Thatcher era of crude monetarism. Many would argue that capitalism's ability to rapidly penetrate post-socialist Europe was a result, at least in part, of ultimate failure of the state socialist system itself (e.g. Potter and Unwin, 1995). However, this 'process' has been overlain by the dynamics of EU enlargement, and its concept of CEE states 'returning to Europe' – an implicitly pejorative understanding that imposition of Russian/Soviet values through communism was not 'of Europe' (despite the obvious irony of Marx and Lenin being firmly European intellectuals, though tempered by the dark days of Stalinism and the fact that Stalin's origins were from the 'barbaric' periphery (Georgia) and thus not 'European').

Fig. 1.4. The way we were. Opposite: (a) The icon of European division and subsequent reunification: looking across the Berlin Wall from West (Wedding, Bernauer Strasse) to East (Schönhauser Allee), July 1975. (b) Celebrating Soviet protection: 30th anniversary poster for the 'liberation' of Czechoslovakia: Prague, August 1975. Above: (c) Mongolia's uniqueness: 'Bypassing capitalism': two posters extolling Mongolia's leap of faith from feudalism straight to communism, Ulan Baatar, September 1984. (d) Celebrating the end of China's 'Cultural Revolution' with portraits of Mao Ce Dung and Hua Guo Feng, Sian, August 1977.

The key institution set up for the specific task of assisting post-communist economic restructuring was the European (sic) Bank for Reconstruction and Development (EBRD), albeit alongside such longstanding global institutions as the World Bank, International Monetary Fund (IMF) and Organization for Economic Cooperation and Development (OECD). The congruence of 'transition' and (particularly European) post-communism was reinforced with the EBRD publishing an annual 'Transition Report', promoted as 'a unique source of information on developments in central and eastern Europe and the Commonwealth of Independent States (CIS)' (EBRD, 2003: 1).

In an expansion of the (capitalist) global economy, the Euro-centric nature of post-communist 'transition' has functioned to assist and to act as a convenient precursor to the process of EU enlargement which has embraced a number of the European post-communist states (Agnew, 2000; Danta and Hall, 2000). Yet the path to global capitalism may be subject to various types of reform processes advancing, stalling or even reversing its progress (Hamilton, 1999: 139). Further, processes advancing capitalism may have constructive or destructive qualities, the results of which may be seen simultaneously in any given locality, region or country. For example:

- Wollmann (2002) talks positively of the 'creative destruction' (a phrase taken from Schumpeter (1934, 1944)) of East German institutions in the transformation of that former state following (re-)unification with West Germany in 1990; while
- Meurs and Ranasinghe (2003) point to the 'de-development' – the loss or reversal of key development structures and institutions – evident in parts of post-communist CEE and much of the FSU during the 1990s.

A recognition of the differential impact of these processes is implicit in the 'Agenda 2000' project which paved the EU accession course, and in the subsequent requirements for each accession state to meet a range of fitness tests. A number of issues have been raised in the process.

- Accommodation of the divergent non-communist, indeed Commonwealth, back-grounds and trajectories of EU entrants Malta and Cyprus, and their impact on the European process, notably the latter with its own particular complications of political division. The process also seems to have difficulty in accommodating the EU candidature of Turkey, for long a staunch NATO ally during the Cold War, and whose position now is also inextricably tied into resolving the Cyprus question (see Chapter 10). Indeed, such has been the inability of the European project to assist a resolution of the Cyprus situation, that the European Council decided, as part of the accession agreement for 'Cyprus', that, in the absence of a settlement, it should suspend application of the *acquis* to the northern part of the island (European Commission, 2003: Chapter 31).
- The previous EU enlargement, the fourth, in 1995, embraced Austria, Finland and Sweden. This so-called 'EFTA enlargement' was notable in that all three entrants had higher per capita levels of gross domestic product (GDP) than the EU average (Edye and Lintner, 1996: 408). According to most measures, the 2004 entrants are significantly poorer and less developed, economically, politically and administratively, than the EU15 members; indeed, the average GDP per capita for the new entrants is only about one-third that of existing member countries (Table 1.2).
- The range of fitness tests applied to the candidate countries, if applied to some existing members, would pose certain challenges and expose inconsistencies. In 1999 an index of EU suitability based solely on economic criteria was developed and applied to all existing and applicant EU countries (Anon., 1999). Belgium was shown to be the most suitable or fittest country to be an EU member, followed by Luxembourg, the Netherlands, Denmark, Portugal, Austria, Ireland, Sweden, France, Spain, Britain, Finland, and Germany in 13th place. The next highest ranked was Slovenia, followed by the Czech Republic, Poland, Cyprus, and then EU founding member Italy. These were followed by Hungary, Malta, Latvia, Estonia, 1981 entrant Greece, Slovakia, Romania,

Fig. 1.5. European icons not (yet) of the EU: (a) Croatia: the Venetian walls of Dubrovnik, World Heritage Site, on the Adriatic. (b) Russia: St Basil's Cathedral on Moscow's Red Square.

Lithuania and Turkey, with Bulgaria finishing in the position of least suitability. The relatively high rating of countries such as Slovenia contrasted with the relatively poor showing of Italy and especially Greece. These findings pointed to the large disparities existing not only in the candidate countries, but also amongst existing EU members. They also underlined the sensitivity of the enlargement process.

Tourism and the Evolving (Capitalist) Global Economy

In certain conceptions of global development processes, such as world systems analysis, all countries are viewed as part of a larger whole, and their trajectories, whether divergent or convergent, are conceived within an holistic global context. The current capitalist global economy, which is seen to have supplanted the political-economic polarities of the Cold War, is viewed by world systems analysts as comprising three basic structural features:

- the world market,
- a multiple-state system, which no single state is able to dominate (a somewhat contestable assumption given contemporary international relations), and
- a three-tier structure of stratification resulting from the relative positioning of societies within, and the relationships between:
  - core processes: consisting of relations which incorporate relatively high wages, advanced technology and a diversified production mix, and
  - peripheral processes: involving low wages, more rudimentary technology and a simple production mix (Taylor and Flint, 2000: 20; Bradshaw, 2001: 29).

The latter dichotomy articulates traditional development concepts of formal and informal economies, which may characterize both the chasm between developed and least developed countries and also the differentiation within developing societies between the formal and informal sectors. Duality within the tourism industries has parallels with such a characterization. In both cases interlinkages between core/formal and peripheral/informal processes may act mutually to reinforce the relative positions.

The world systems approach argues that the position of a particular state in the world economy is determined by the mix of 'core' and 'peripheral' processes and the relative domination of one of these conditions. Where particular, dynamic mixes of both occur, world systems analysis refers to a third, 'semi-periphery' condition, which is particularly characteristic of two different types of countries: those resource-exporting and those newly industrializing (Knox and Agnew, 1998: 62). The old Soviet bloc was seen to fall into this category (Bradshaw and Lynn, 1994), and during the 1990s this view persisted, albeit with the less developed former Soviet republics (Armenia, Azerbaijan, Georgia, Kazakhzstan Kyrgyzstan, Tajikistan, Turkmenistan and Uzbekistan) being consigned to the 'periphery' (Knox and Agnew, 1998). Indeed, all of the societies considered in this volume were regarded as semi-peripheral or peripheral.

Further, the majority of states of the former Soviet Union remain disengaged from the global economy, and while EU membership will assist CEE states to join the global core – including the Baltic trio which were unwilling republics of the Soviet Union for half a century – the southern regions of the former USSR and particularly the land-locked Central Asian states, dominated by peripheral processes, appear destined to remain within the global periphery for some time to come (Bradshaw, 2001: 42–43).

Nevertheless, a notable characterization of international tourism development is that the process is subject to considerable regional variation, with new centres continually arising. Peripheries may emerge as centres or may lose their earlier appeal (Harrison, 2001: 40).

The trajectories of the Mediterranean EU accession states – Malta and Cyprus – of course derived from very different points of origin compared to their CEE counterparts. Indeed, the term 'transition' has been only rarely applied to these societies, and it falls to a French author (Miossec, 1994) to provide an early published reference to Malta as being 'in transition': in this case viewed in terms of its demographic and economic characteristics.

References in the literature to South Africa's 'transition' (e.g. Mills, 2002; von Holdt, 2002) relate particularly to the political dimension and to the post-apartheid context for social and economic change. They often focus on specific institutions of change such as the Truth and Reconciliation Committee (TRC) of 1996/97 (e.g. Buur, 2003). Despite being by far the most visited country in Africa, attempts to capture this change and to articulate it for tourism and leisure purposes have not met with wholesale success within a still fractured and unsure society (e.g. Witz et al., 2001; Crampton, 2003).

Fig. 1.6. South Africa: long-distance bus within Johannesburg's landscape of inequality.

Although substantial political change has not come about in China and Vietnam, both are often characterized in the literature as having 'transitional' economies (e.g. Lloyd, 2003; Zheng, 2003; Zhou *et al*., 2003), often explicitly encompassing social change (e.g. Luttrell, 2001; Tsui, 2002). In Cambodia, Roberts (2002) suggests that political transition has not accorded with accepted Western models, ironically because of the way in which elections were imposed by external agencies. By contrast, Fritz (2002) argues that Mongolia has contradicted the assumptions of modernization theory and has accepted 'modes of transition' by retaining its political elites yet proceeding to consolidate democracy, in this case positively aided by the promotion of external agencies. The relative roles and variable relations between endogamous and exogenous processes, as suggested here, is a theme which recurs in this volume.

Analysts of Cuban development may carry a little more ideological baggage than others, and in the early years of the 21st century a literature has arisen, notably from the exile community in Florida, employing the term 'transition' as wish-fulfilment (e.g. Otero and O'Bryan, 2002; Miller, 2003), as the end of Fidel Castro's 'maximum leader' role draws ever closer. But, as briefly discussed in Chapter 2, since the collapse of Soviet support, Cuba

has pursued a degree of economic reform and restructuring, and this in turn has exerted a significant influence on the country's tourism trajectory.

Fig. 1.7. China: the heavily visited Badaling section of the Great Wall, World Heritage Site.

Fig. 1.8. Vietnam: staged authenticity for the tourist gaze in Ho Chi Minh City.

Fig. 1.9. Cuba: communist iconography as tourism attraction. Re-enacting the taking of Santa Clara (and specifically the tourist hotel) during the 1959 revolution.

With particular reference to the spatial impact of extension of the global economy to post-communist CEE and FSU, the peoples and territories here are experiencing various types and degrees of transformation through the interaction of three key components, the first being strongly constrained or enhanced in different countries, regions and localities by the nature and interaction of the other two:

- progress from a predominantly centrally managed, state-owned (command-type) economic system designed to function with a single-party communist political system, towards a more privatized market-type economy managed by a myriad of decision makers within a more democratic, civil society and new regulatory frameworks;

- legacies and impacts of structures, institutions, networks, behaviours and mind-sets which have been inherited from the preceding state socialist era, as well as those more deeply 'embedded' from pre-socialist times; and
- the human, organizational and institutional responses at various geographic scales to the opportunities created, or to the constraints and pressures imposed, by the transformation process itself (Hamilton, 1999: 136).

Thus while transition from 'communist' to 'post-communist' economies concerns macro-economic shifts: (a) from accumulation through state-led industrialization to accumulation through commodification and export orientation, and (b) from redistribution to 'marketization', this is not the case for other societies examined in the book which are differently influenced in their trajectories by capitalist expansion. Indeed, by contrast, those 'transitional' societies explicitly not addressed in this volume – advanced market economies – might be portrayed as shifting: (a) from economies of scale to economies of scope, and (b) from Keynesian welfare state economics to neo-liberalism and post-Keynesian workfare approaches. But this in itself is a gross over-simplification of the changes in advanced capitalism (e.g. see Goodin, 2002; Gray, 2002; Peck, 2002), the details of which and their interrelationships with tourism development being well outside the scope of the current work.

As an integral part of the global expansion of capitalism, tourism may be seen as a concomitant and often a prime example of: (a) growth, privatization and 'flexibilization' of service industries; (b) reduction of centralization, subsidy and bureaucratic control; (c) emphasis on private initiative and individual entrepreneurial activity, not least in the encouragement of SMEs; (d) emphasis, through comparative advantage, on niche specialization and sector segmentation; (e) emphasis of a sectoral shift from secondary to tertiary activities, from manufacturing to services; (f) exposure of enterprises to national and international market forces; and (g) market penetration and FDI of transnational corporations. The nature and

impact of such processes are moulded by three dynamic contexts (e.g. Johnson, 1997) which involve changes in:

- individual countries' political economy;
- a country's pattern of external relationships; and
- the world tourism industry itself.

During the 1980s and 1990s evolving symbiotic relationships between tourism development and processes of economic, political and social restructuring could be identified. Worthington (2001), for example, argued that the service sector including tourism was the transition catalyst for Estonia. Certainly in the case of the CEECs at least, this embraced issues of price liberalization, encouragement of entrepreneurial activity, large- and small-scale privatization, deregulation, divestment, changing circumstances for personal mobility, enhanced and reoriented foreign trade and investment (e.g. Kilvits and Purju, 2003), including outward FDI (Kalotay, 2003), and currency convertibility (Hall, 1991: 11–12; Jaakson, 1996).

The infrastructural sub-text of developing adequate transport and communications to support tourism and wider regional development processes has been a critical element of the wider continental process of realignment and re-integration in Europe. Although tourism's specific contribution is variable and may not necessarily be readily apparent, major regional and international transport infrastructures continue to be upgraded as part of a wider process of improved access from, and closer integration with, global markets (Hall, 1993a,b; Lijewski, 1996; Taylor, 1998). Yet, domestic core–periphery relationships may be exacerbated both by the growth of hub and spoke transport systems and by an emphasis on prestige international infrastructure at the expense of services for domestic localities. In the latter case, for example, the development of high-speed rail connections within and into Poland has been accompanied by a decline in the country's remaining rail services (Kossowski, 2002).

Achievement of currency convertibility became one symbol of the fiscal maturity of post-socialist states and of their integration into the capitalist global economy. While some states have achieved convertibility, the currencies of others remain inconvertible and thus a pejorative indicator of economic status. Others again have formally adopted the euro. Economic restructuring programmes, whether of the 'shock therapy' or more gradual variety, have been a part of and/or have needed to respond to a range of previously unknown economic circumstances: loss of former markets, industrial collapse and unemployment, price inflation and diminishing living standards. In attempting to combat these circumstances, trade balance improvement has been a vital governmental objective, although the actual nature and acknowledgement of tourism's role in this process has varied considerably between countries, regions and localities.

By encouraging greater and closer interaction between formerly restricted host populations and the outside world, both inbound and outbound tourism may be seen as a catalyst of change, whether as a positive educational force or in setting up potentially negative demonstration effects. Integration into the global economy, and for several European states preparation for EU accession, has been a prime foreign-policy driving force. The ten candidates for the 2004 enlargement have revealed mixed tourism fortunes in recent years (Table 1.2), although data for the Baltic states show significant growth in receipts during the second half of the 1990s (Table 1.3). Generally, the more stable countries of postcommunist Central Europe (notably the Czech Republic, Hungary and Poland) – those spatially and structurally best placed for EU integration and exhibiting the most explicit characteristics of global incorporation – experienced the most notable early post-communist growth in international tourism activity, although this was partly derived from processes already under way before 1989 (for example, see Chapter 5 later in this volume). Tourism and its economic impact in relatively unstable south-eastern Europe and the CIS has stagnated or even declined (Hall, 1998a,b, 2000a,b; Light and Dumbrăveanu, 1999) (for example, see Chapter 7 in this volume). In these countries, where private sector underfunding often poses major challenges, infrastructural shortcomings have persisted (Bachvarov, 1999; Holland, 2000).

Such outcomes reflect wider patterns of divergence in economic development prospects between Central Europe and the Baltic states on the one hand and SEE and the CIS on the other. Hamilton (1999: 139) argued fairly starkly that contemporary economic divergence is an embodiment of:

- path dependencies (see Chapter 2) embedded in the temporal and contextual differences between 70 years of state socialism in much of the FSU (though this does not account for SEE's position, largely derived from a legacy of limited development), and around 40 years in Central Europe and the Baltic states;
- the more clear-cut rejection of state socialism in Central Europe and the Baltic states which stems from its imposition on them by the Soviet imperial power in the 1940s;
- the ability to keep alive or, in the latter stages of communism, reinvigorate entrepreneurship. By way of contrast, it could be argued that pre-communist SEE had not developed a strong entrepreneurial sector (nor a functioning civil society, in some cases) and was therefore unable to maintain or rejuvenate such a legacy;
- that purged of Christian or liberal values, it was not considered morally wrong in the USSR to exploit any system – socialist, capitalist or 'transitional' – by using personal contact networks, stealing, and other 'informal modes' of redistribution.

Certainly, in political terms, the Czech Republic, Hungary and Poland managed early in the post-communist process to develop an ethos of unity in diversity, reinforced by relative consensus among political and economic elites – not least the aspiration for EU membership – which brought relative stability. By contrast, in Belarus, Serbia and Ukraine, for example, elite disunity deepened rather than abated (Highley and Pakulski, 2000).

The Following Chapters

In this volume, critical evaluation and exemplification of tourism and 'transition' embraces contributions focusing on Hungary, Poland, Serbia, Estonia, Romania and Central and

Table 1.2. Selected countries in transition: international tourist arrivals, 1990–2000.

Country	Series	International tourist arrivals (in millions)					Average annual growth 1995–1999
		1990	1995	1998	1999	2000	
CEE Post-Communist 2004 EU entrants							
Czech Republic	TCE	7.28	3.38	5.48	5.61	5.70	13.5
Estonia	TF	na	0.53	0.83	0.95	1.10	15.7
Hungary	VF	20.51	19.62	16.81	14.40	15.57	−7.4
Latvia	TF	na	0.52	0.57	0.49	na	−1.6
Lithuania	TF	na	0.65	1.42	1.42	1.23	21.6
Poland	TF	3.40	19.22	18.78	17.95	17.40	−1.7
Slovakia	TCE	0.82	0.90	0.90	0.98	1.05	1.9
Slovenia	TCE	0.65	0.73	0.98	0.88	1.09	4.8
Mediterranean 2004 EU entrants (including northern Cyprus)							
Cyprus	TF	1.56	2.10	2.22	2.43	2.69	3.8
Malta	TF	0.87	1.12	1.18	1.21	1.22	2.1
Other CEE (including Turkey)							
Albania	THS	0.03	0.04	0.03	0.04	na	−0.6
Belarus	TF	na	0.16	0.36	na	na	na
Bosnia-Hercegovina	TF	na	0.04	0.09	0.09	0.11	24.5
Bulgaria	TF	1.59	3.47	2.67	2.47	2.79	−8.1
Croatia	TCE	7.05	1.49	4.50	3.81	5.83	26.5
Georgia	TF	na	0.09	0.32	0.38	na	45.8
Macedonia, FYR	TCE	0.56	0.15	0.16	0.18	0.22	5.3
Moldova	TF	na	0.03	0.02	0.01	0.02	−18.7
Romania	TF	3.01	2.76	2.97	3.21	3.27	3.9
Russia	VF	na	10.29	15.81	18.50	21.17	15.8
Serbia/Montenegro	TCE	1.19	0.23	0.28	0.15	na	−9.6
Turkey	TF	4.80	7.08	8.96	6.89	9.59	−0.7
Ukraine	TF	na	3.72	6.21	4.23	na	3.3
Other Former Soviet Union							
Armenia	TCE	na	0.01	0.03	0.04	0.03	36.0
Azerbaijan	TF	na	0.09	0.48	0.60	0.68	59.5
Kazakhstan	–	na	na	na	na	na	na
Kyrgyzstan	TF	na	0.04	0.06	0.07	na	17.7
Tajikistan	TF	na	na	0.51	na	na	na
Turkmenistan	TF	na	0.22	0.30	na	na	na
Uzbekistan	VF	na	0.09	0.27	na	na	na
USSR		7.20	–	–	–	–	–
Other Asian							
Cambodia	TF	0.02	0.22	0.29	0.37	0.47	13.7
China	TF	10.48	20.03	25.07	27.05	31.23	7.8
Korea, DPR	na	0.12	0.13	0.13	na	na	na
Lao PDR	TF	0.01	0.06	0.20	0.26	0.30	44.1
Mongolia	TF	0.15	0.10	0.20	0.16	0.16	10.2
Vietnam	VF	0.25	1.35	1.52	1.78	2.14	7.2
Post-apartheid							
South Africa	VF	1.03	4.68	5.90	6.03	6.00	6.5
Latin Caribbean							
Cuba	TF-air	0.33	0.74	1.39	1.56	1.70	20.4

Key: TCE, tourist arrivals at all accommodation establishments; TF, tourist arrivals at frontiers;
THS, tourist arrivals at hotels; VF, visitor arrivals at frontiers; *na*, data not available.
Sources: WTO, 2002a,b.

Table 1.3. Selected countries in transition: international tourist receipts, 1990–2000.

Country	International tourist receipts (in US$ millions)					Average annual growth 1995–1999
	1990	1995	1998	1999	2000	
CEE Post-Communist 2004 EU entrants						
Czech Republic	419	2,875	3,719	3,035	2,869	1.4
Estonia	na	353	534	560	505	12.2
Hungary	824	2,640	3,514	3,394	3,424	6.5
Latvia	na	20	182	118	131	55.9
Lithuania	na	77	460	550	391	63.5
Poland	358	6,614	7,946	6,100	6,100	−2.0
Slovakia	70	620	489	461	432	−7.1
Slovenia	721	1,084	1,088	954	957	−3.1
Mediterranean 2004 EU entrants (including Northern Cyprus)						
Cyprus	1,258	1,788	1,696	1,878	1,894	1.2
Malta	496	660	661	675	650	0.6
Other CEE (including Turkey)						
Albania	4	65	54	211	na	34.2
Belarus	na	23	22	13	17	−12.8
Bosnia-Hercegovina	na	7	21	21	17	31.6
Bulgaria	320	473	966	932	1,074	18.5
Croatia	1,704	1,349	2,733	2,493	2,758	16.6
Georgia	na	na	423	400	na	na
Macedonia, FYR	45	19	15	40	37	20.5
Moldova	na	4	4	2	4	−15.9
Romania	106	590	260	254	364	−19.0
Russia	na	4,312	6,508	7,510	na	14.9
Serbia/Montenegro	419	42	35	17	na	−20.2
Turkey	3,225	4,957	7,177	5,203	7,636	1.2
Ukraine	na	3,865	3,317	2,124	na	−13.9
Other Former Soviet Union						
Armenia	na	5	10	27	45	52.4
Azerbaijan	na	70	125	81	na	3.7
Kazakhstan	na	122	407	363	na	31.3
Kyrgyzstan	na	5	8	na	na	na
Tajikistan	na	na	na	na	na	na
Turkmenistan	na	na	192	na	na	na
Uzbekistan	na	na	21	na	na	na
USSR	2,752	–	–	–	–	–
Other Asian						
Cambodia	na	100	166	190	228	17.4
China	2,218	8,733	12,602	14,098	16,231	12.7
Korea, DPR	na	na	na	na	na	na
Lao PDR	3	25	80	97	114	40.3
Mongolia	5	21	33	28	na	7.5
Vietnam	85	86	86	na	na	na
Post-apartheid						
South Africa	992	2,125	2,738	2,526	na	4.4
Latin Caribbean						
Cuba	243	977	1,571	1,714	1,756	15.1

Key: *na*, data not available.
Sources: WTO, 2002a,b.

Eastern Europe more generally, Malta, Cyprus and the eastern Mediterranean, Kyrgyzstan, China and South Africa. It is therefore far from being a geographically comprehensive review of tourism within transforming countries, however defined. Rather, the following chapters attempt to emphasize and articulate a number of key themes and frameworks which may help us better understand the many contemporary and simultaneous local and global variables, both influencing and influenced by the processes and structures of tourism development in transforming societies, during the first decade of the 21st century.

The range of chapters (3–8) based on contemporary post-communist experience in Central and Eastern Europe and the former Soviet Union can be seen to reflect, in turn, optimism, pessimism and paradox, in terms of trends, relationships and impacts. Drawing on research and consultancy experience from Central and Eastern Europe, in Chapter 3, Lesley Roberts introduces themes subsequently echoed in a number of chapters, by highlighting the importance of social capital for rural areas where the highly localized nature of tourism is most evident. Emphasizing that the locality – including local people – provides the services produced and delivered and the experiences created, an endogenous base for development is viewed as being most likely to provide a sustainable means of regional rural development. But as indicated subsequently from Estonia (Chapter 6) and Kyrgyzstan (Chapter 8), there can be substantial problems and paradoxes attached to this.

Chapter 4 sees Wacław Kotlinski drawing out themes of under-investment and declining tourist numbers in Poland. Crucially, he identifies a growing gap between income derived from tourism to support state budgets and actual national expenditure for tourism infrastructure and development, and offers recommendations for state action to stimulate tourism and enhance the recognition of its economic role within the country. This would seem to be particularly important given the boost to tourism Poland's EU entry is likely to stimulate. In Chapter 5, Zsuzsanna Behringer and Kornélia Kiss detail and evaluate the apparently rising level and nature of foreign direct investment in Hungary's tourism industry in a national context where international tourist arrival numbers have been declining since 1995, and where *per capita* spending is still relatively low. Such a paradox would appear, superficially, difficult to reconcile. To what extent will this be ameliorated or exacerbated by Hungary's accession to the European Union? In the face of the disastrous collapse of Serbia's tourism industry after 1989, Jovan Popesku and Derek Hall argue in Chapter 7 that coordinated sustainable development based on natural resources with planned, limited growth, represents the only reasonable forward strategy for the reconstruction of an international tourism industry in land-locked Serbia, where any likely EU membership is still some way off.

The chapter (6) by Barry Worthington raises interesting paradoxes concerning the endogamous–exogenous dimension when addressing relationships between national heritage and tourism in the context of post-Soviet Estonia. Within the Soviet Union, the Estonians' 'minority' culture was commodified for passive consumption, yet Estonians themselves regarded their heritage as dynamic and participatory, and employed it subversively to re-create a civil society as an alternative to Soviet institutions. However, according to Worthington's analysis, regaining independence in 1991 removed this imperative, and the harnessing of heritage for tourism remained perceived in pejorative terms especially as (Western) tour companies continued to echo Soviet practice in the depiction of Estonian heritage. How will Estonia's EU accession change this? In the case of a second former Soviet territory with contrasting cultural, economic and environmental characteristics, Peter Schofield argues in Chapter 8 that Kyrgyzstan's tourism product has been poorly positioned and the development and projection of an appropriate image is critically required to address this. Although cultural icons are available to assist in raising awareness of identity and distinctiveness, Schofield points to a lack of capital resources and political instability as continuing constraints on an adequate response to this need.

Turning to the Mediterranean basin, Maria Attard and Derek Hall evaluate some of the impacts of transition towards EU accession for Malta's tourism and transport sectors in Chapter 9. New brand images and marketing programmes, coupled with an emphasis on high value niche development, have been outward signs of a restructuring of the Maltese tourism product. While physical replacement of ancient public service buses – regarded as a heritage attraction and experience in their own right – is part of the accession agenda, potentially negative images are reflected in growing traffic congestion and pollution and the long-overdue need for a comprehensive upgrading of transport infrastructure. The interplay of these factors is underscored by an uncertain and divisive domestic political context.

Division is certainly a salient theme for Habib Alipour and Hasan Kilic's appraisal of the context in which tourism in northern Cyprus has been developed since that island's political partition in 1974 (Chapter 10). The focus of the chapter is inadequate tourism governance, the reasons for which vary – governments often lack the will to implement policies, tourism plans may be narrowly defined and implemented without reference to wider development strategy or the state may be incapable of implementing policy due to inefficiency and/or corruption – although the expansion of tourism activity may trigger government action (Harrison, 2001: 38–39). In the case of the Turkish Republic of North Cyprus, the authors offer a number of reasons why institutions overseeing tourism development have failed to address, define and formulate clear policy, which has resulted in an absence of integrated development. They point to the need to restructure institutions and to establish an appropriate legislative framework supported by confidence-building measures to attract investment and increase the perpetually low numbers of tourists. Such measures will reflect the significance of both internal and external factors. The path of EU accession for 'Cyprus' and Turkey may prove to be a significant political and economic catalyst in this respect.

Global tourism experienced exceptional years in 2000 and 2001. In 2000 international tourism grew by 45 million arrivals, while in the following year international arrivals declined by 0.6%, the first year of negative growth since 1982. Of course, the events of

11 September 2001 greatly influenced the year's final figures, particularly adversely affecting American destinations and certain sectors such as the ocean cruise market (Hall, 2004). A year later, in October 2002, bombings at tourist attractions in Bali helped to sustain some negative trends (WTO, 2002b: 2). Providing a strong methodological emphasis, in Chapter 11 Antonis Theocharous argues the need for assessment, evaluation and analysis of the interrelationships between such destabilizing events and the tourism industry, in relation to four eastern Mediterranean destination countries. His neural networks application provides a useful approach to tourism demand modelling and a spur to further research in this little-explored field.

In a relatively short time China has become the 'unrivalled leader of Asian tourism' (WTO, 2002b: 4). Themes of economic liberalization within a still relatively rigid state political-bureaucratic framework permeate the two chapters looking at Chinese experience. Rong Huang in Chapter 12 identifies and evaluates the roles played by the Hunan provincial government in developing its international and domestic tourism development policies. Placed within the wider context of China's tourism development phases, the author foresees new roles for the provincial government to play and makes a number of recommendations for responding to the challenges facing the development of Hunan tourism as it draws further away from strategy based on a centrally planned economy.

Issues relating to the tensions between the local and the global, and endogenous and exogenous development factors, are emphasized in the chapter (13) by Takayoshi Yamamura, who looks at the World Heritage Site (WHS) of the old town of Lijiang, in an ethnic minority area of Yunnan Province. Until the later 1970s, when tourism was severely restricted, handicrafts and traditions of ethnic minorities were largely suppressed. An 'open-door' policy from 1978 increasingly involved foreign capital (Zhang et al., 1999), and allowed a revival of ethnic minority religions and traditions which were portrayed to tourists as examples of the diversity of Chinese culture (Sofield and Li, 1998: 369–376). Previous research (Swain, 1990) has emphasized how ethnic groups may be differentially placed to take advantage of their newly-discovered tourism roles in response to the Chinese government's 'commoditizing' such ethnicity (Harrison, 2001: 38–39). Following an earthquake in 1996 and since designation by UNESCO as a WHS in 1997, there has been a dramatic increase in tourism businesses in Lijiang, largely driven by an influx of Chinese Han majority peoples from outside of the region selling goods largely devoid of local character. This situation is exacerbated by the fact that the local Naxi people tend to lack business and management know-how and cannot draw upon government support policies. Yamamura argues that there would appear to be a need for policies to promote local entrepreneurial endeavour through support for indigenous organizations and networks to help stimulate high added value goods and services which can draw upon and illuminate local culture and heritage.

Finally, in relation to a country in transformation following its own particular restructuring processes of the 1990s, in Chapter 14 Jenny Briedenhann and Steve Butts argue that a tourism boom, projected by government as a panacea for the country's economic ills, has not materialized in South Africa. The country is by far the most important international tourism destination in the whole of Africa, receiving a fifth of the continent's arrivals and one quarter of its international tourism receipts (WTO, 2002a,b). Yet, the absence of a national funding strategy for tourism development has resulted in unilateral action by provinces, with negative consequences for national development coherence, public–private partnerships and industry integration. Crucially, the authors contend that a coherent national organizational structure, with a clear delineation of functions and responsibilities, continuously monitored, is urgently required.

The issues, problems, paradoxes and opportunities presented in these chapters are diverse and pursued from different perspectives, yet they share common themes of tourism within processes of transformation. They also emphasize the important influences of, and the need for tourism analysts to better appreciate, underlying political cultures, their inheritances and influences – both formal and informal – on social, economic and environmental transformation. Such transformation is a key influence on, and component of, contemporary local and global tourism trajectories. This theme is taken up and elaborated in the next chapter.

References

Agnew, J. (2000) How many Europes? In: Hall, D. and Danta, D. (eds) *Europe Goes East: EU Enlargement, Diversity and Uncertainty*. The Stationery Office, London, pp. 45–54.

Anon. (1999) EU enlargement. *The Economist*, 18 December, p. 148.

Bachvarov, M. (1999) Troubled sustainability: Bulgarian seaside resorts. *Tourism Geographies* 1(2), 192–203.

Blonski, K. (1998) *Krakow 2000: European City of Culture*. City and Voivodship of Kraków, Kraków.

Bradshaw, M. (2001) The post-socialist states in the world economy: transformation trajectories. *Geopolitics* 6(1), 27–46.

Bradshaw, M.J. and Lynn, N.J. (1994) After the Soviet Union: the post-Soviet states in the world system. *Professional Geographer* 46(4), 439–449.

Brown, F. (2004) The final frontier? Tourism and space. *Tourism Recreation Research* 29(1).

Buchroithner, M. (2002) Creating the virtual Eiger North Face. *Journal of Photogrammetry and Remote Sensing* 57(1), 114–125.

Buur, L. (2003) 'In the name of the victims': the politics of compensation in the work of the South African Truth and Reconciliation Commission. In: Gready, P. (ed.) *Political Transition: Politics and Cultures*. Pluto Press, London, pp. 148–164.

Cheong, R. (1995) The virtual threat to travel and tourism. *Tourism Management* 16(6), 417–422.

CIA (Central Intelligence Agency) (2003) *The World Factbook 2002*. CIA, Washington, DC <http://www.cia.gov/cia/publications/factbook/>.

Crampton, A. (2003) The art of nation-building: (re)presenting political transition at the South African national gallery. *Cultural Geographies* 10(2), 218–242.

Danta, D. and Hall, D. (2000) Introduction. In: Hall, D. and Danta, D. (eds) *Europe Goes East: EU Enlargement, Diversity and Uncertainty*. The Stationery Office, London, pp. 3–14.

EBRD (European Bank for Reconstruction and Development) (2003) *Transition Report*. EBRD, London <http://www.ebrd.com/pubs/index.htm>.

Edye, D. and Lintner, V. (1996) Conclusion: prospects for the new Europe. In: Edye, D. and Lintner, V. (eds) *Contemporary Europe*. Prentice Hall, Hemel Hempstead, UK, pp. 393–411.

European Commission, The (2003) *Report on the Results of the Negotiations on the Accession of Cyprus, Malta, Hungary, Poland, the Slovak Republic, Latvia, Estonia, Lithuania, the Czech Republic and Slovenia to the European Union*. European Commission, Brussels <http://www.europa.eu.int/ enlargment/negotiations/pdf/negotiations_report_to_ep.pdf>.

Fritz, V. (2002) Mongolia: dependent democratization. *Journal of Communist Studies and Transition Politics* 18(4), 75–100.

Glowka, L. (2003) Putting marine scientific research on a sustainable footing at hydrothermal vents. *Marine Policy* 27(4), 303–312.

Goodin, R.E. (2002) Structures of mutual obligation. *Journal of Social Policy* 31(4), 579–596.

Gray, A. (2002) European perspectives on welfare reform: a tale of two vicious circles? *European Societies* 4(4), 359–380.

Gready, P. (2003) Introduction. In: Gready, P. (ed.) *Political Transition: Politics and Cultures*. Pluto Press, London, pp. 1–26.

Hall, D. (1991) Introduction. In: Hall, D. (ed.) *Tourism and Economic Development in Eastern Europe and the Soviet Union*. Belhaven Press, London, and Halsted Press, New York, pp. 3–28.

Hall, D. (1993a) Impacts of economic and political transition on the transport geography of Central and Eastern Europe. *Journal of Transport Geography* 1(1), 20–35.

Hall, D. (1993b) *Transport and Economic Development in the New Central and Eastern Europe*. Belhaven Press, London, and Halsted Press, New York.

Hall, D. (1998a) Central and Eastern Europe. In: Williams, A.M. and Shaw, G. (eds) *Tourism and Economic Development in Europe*. John Wiley & Sons, Chichester, UK, pp. 345–373.

Hall, D. (1998b) Tourism development and sustainability issues in Central and South-eastern Europe. *Tourism Management* 19(5), 423–431.

Hall, D. (2000a) Evaluating the tourism–environment relationship: Central and Eastern European experiences. *Environment and Planning B: Planning and Design* 27(3), 411–421.

Hall, D. (2000b) Tourism as sustainable development? The Albanian experience of 'transition'. *International Journal of Tourism Research* 2(1), 31–46.

Hall, D. (2004) Ocean cruising: market dynamics, product responses and onshore impacts. In: Pinder, D. and Slack, B. (eds) *Shipping and Ports in the 21st Century*. Routledge, London.

Hamilton, F.E.I. (1999) Transformation and space in Central and Eastern Europe. *Geographical Journal* 165(2), 135–144.

Harrison, D. (2001) Tourism and less developed countries: key issues. In: Harrison, D. (ed.) *Tourism in the Less Developed World: Issues and Case Studies*. CAB International, Wallingford, UK, pp. 23–46.

Highley, J. and Pakulski, J. (2000) Jeux de pouvoir élites et consolidation de la démocratie en Europe centrale et orientale. *Revue Française de Science Politique* 50(4-5), 657–678.

Holland, J. (2000) Consensus and conflict: the socioeconomic challenge facing sustainable tourism development in Southern Albania. *Journal of Sustainable Tourism* 8(5), 510–524.

Jaakson, R. (1996) Tourism in transition in post-Soviet Estonia. *Annals of Tourism Research* 23(3), 617–634.

Johnson, M. (1997) Hungary's hotel industry in transition, 1960–1996. *Tourism Management* 18(7), 441–452.

Kalotay, K. (2003) Outward foreign direct investment from economies in transition in a global context. *Journal for East European Management Studies* 8(1), 6–24.

Kilvits, K. and Purju, A. (2003) The outward foreign direct investment from the Baltic States. *Journal for East European Management Studies* 8(1), 83–95.

Klein, S. and Loebbecke, C. (2003) Emerging pricing strategies on the web: lessons from the airline industry. *Electronic Markets* 13(1), 46–58.

Knox, P. and Agnew, J.A. (1998) *The Geography of the World Economy*. Arnold, London.

Kossowski, T. (2002) Szybkie polaczenia kolejowe w Polsce i ich zmiany w latach 1975–1999. *Przeglad Geograficzny* 74(2), 229–242.

Liebman Parrinello, G. (2001) The technological body in tourism research and praxis. *International Sociology* 16(2), 205–219.

Light, D. and Dumbrăveanu, D. (1999) Romanian tourism in the post-communist period. *Annals of Tourism Research* 26(4), 898–927.

Lijewski, T. (1996) The impact of political changes on transport in Central and Eastern Europe. *Transport Reviews* 16, 37–53.

Lloyd, K. (2003) Contesting control in transitional Vietnam: the development and regulation of traveller cafés in Hanoi and Ho Chi Minh City. *Tourism Geographies* 5(3), 350–366.

Luttrell, C. (2001) Institutional change and natural resource use in coastal Vietnam. *GeoJournal* 55(2-4), 529–540.

Meurs, M. and Ranasinghe, R. (2003) De-development in post-socialism: conceptual and measurement issues. *Politics and Society* 31(1), 31–54.

Miller, N. (2003) The absolution of history: uses of the past in Castro's Cuba. *Journal of Contemporary History* 38(1), 147–162.

Mills, G. (2002) Chile and South Africa – lessons and opportunities from political and economic transition. *Unisa Latin American Report* 18(2), 4–14.

Miossec, J.M. (1994) Malte en transition: démographie, économie et gestion de l'espace. *Revue du Monde Musulman et de la Méditerranée* 71, 199–216.

Otero, G. and O'Bryan, J. (2002) Cuba in transition? Civil society's challenge to the Castro regime. *Latin American Politics and Society* 44(4), 29–57.

Peck, J. (2002) Political economies of scale: fast policy, interscalar relations, and neoliberal workfare. *Economic Geography* 78(3), 331–360.

Potter, R. and Unwin, T. (1995) Urban–rural interaction: physical form and political processes. *Third World Cities* 12(1), 67–73.

Roberts, D. (2002) Political transition and elite discourse in Cambodia, 1991–99. *Journal of Communist Studies and Transition Politics* 18(4), 101–118.

Schumpeter, J. (1934) *The Theory of Economic Development*. Oxford University Press, Oxford.

Schumpeter, J. (1944) *Capitalism, Socialism and Democracy*. George Allen and Unwin, London.

Simpson, F.R. (1999) Tourist impact in the historic centre of Prague: resident and visitor perceptions of the historic built environment. *Geographical Journal* 165(2), 173–183.

Sofield, T.H.B. and Li, F.M.S. (1998) Tourism development and cultural policies in China. *Annals of Tourism Research* 25(2), 362–392.

Swain, M.B. (1990) Commoditizing ethnicity in Southwest China. *Cultural Survival Quarterly* 14(1), 26–30.

Taylor, P.J. and Flint, C. (2000) *Political Geography: World Economy, Nation-state and Locality*. Prentice-Hall, Harlow, UK.

Taylor, Z. (1998) Polish transport policy: an evaluation of the 1994/5 White Paper. *Journal of Transport Geography* 6(3), 227–236.

Tsui, M. (2002) Managing transition: unemployment and job hunting in urban China. *Pacific Affairs* 75(4), 515–534.

UNECE (United Nations Economic Commission for Europe) (2003) *Trends in Europe and North America*. UNECE, Geneva <http://www.unece.org/stats/trend/trend _h.htm>.

United Nations (2003a) *Millennium Indicators Database*. UN Department of Economic and Social Affairs, New York <http://millenniumindicators.un.org/unsd/>.

United Nations (2003b) *Statistics Division*. UN Department of Economic and Social Affairs, New York <http://unstats.un.org/unsd/>.

Urry, J. (1995) *Consuming Places*. Routledge, London.

von Holdt, K. (2002) Social movement unionism: the case of South Africa. *Work, Employment and Society* 16(2), 283–304.

Williams, P. and Hobson, J.S.P. (1995) Virtual reality and tourism: fact or fantasy. *Tourism Management* 16(6), 423–427.

Williamson, M. (2003) Space ethics and protection of the space environment. *Space Policy* 19(1), 47–52.

Witz, L., Rassool, C. and Minkley, G. (2001) Repackaging the past for South African tourism. *Daedalus* 130(1), 277–296.

Wollmann, H. (2002) Local government and politics in East Germany. *German Politics* 11(3), 153–178.

World Bank, The (2003) *World Development Indicators Database*. World Bank, New York <http://www.worldbank.org/data/wdi2003/index.htm>.

Worthington, B. (2001) Riding the 'J' curve – tourism and successful transition in Estonia? *Post-Communist Economies* 13(3), 389–400.

WTO (World Tourism Organization) (2002a) *Compendium of Tourism Statistics 1996–2000*. WTO, Madrid.

WTO (World Tourism Organization) (2002b) *Tourism Highlights 2002*. WTO, Madrid <http://www.world-tourism.org>.

Zhang, H.Q., Chong, K. and Ap, J. (1999) An analysis of tourism policy development in modern China. *Tourism Management* 20(4), 471–485.

Zheng, X.P. (2003) A note on the equilibrium and optimum of a transitional urban economy: the case of China. *Papers in Regional Science* 82(1), 137–146.

Zhou, X., Zhao, W., Li, Q. and Cai, H. (2003) Embeddedness and contractual relationships in China's transitional economy. *American Sociological Review* 68(1), 75–102.

2 Key themes and frameworks

Derek Hall

This chapter outlines and discusses three key themes which can provide frameworks for this volume's attempt to evaluate and exemplify the nature and roles of, and influences on, tourism development in contemporary processes of 'transition'. These key themes are:

- the conceptual value of 'transition';
- the processes of 'transition' towards EU accession; and
- the definition/redefinition of identity through tourism.

In so doing, this chapter has two major objectives:

- to raise salient issues concerning the conceptualization of tourism within transition; and
- to bring analytical coherence to an understanding of the change processes being experienced in the range of 'transitional' societies featured in this volume.

The structure and content of this chapter draw on a number of sources and influences, one being Dingsdale's (1999) polemic on the direction of 'geographies of post-socialist Europe' and his identification of four paradigms bringing order to the geographical interpretation of such a Europe. Heavily adapted interpretations of three of those paradigms help to structure this chapter, while aspects of the fourth were addressed in Chapter 1.

Tourism within 'Transition': Some Conceptual Considerations

As noted in Chapter 1, use of the term 'transition' has become almost synonymous with prescriptive forms of post-communist restructuring processes, despite: (a) the range of other types of 'transitions' long conceptualized, such as demographic; (b) societies experiencing political and/or economic transformation but not from a state socialist condition, such as South Africa; and (c) state socialist societies pursuing economic but not political change, such as China. Likewise, tourism's role in relation to transition has tended to be viewed within the same narrow and somewhat prescriptive parameters. One of the purposes of this volume is to help broaden the debate from which these handicaps derive.

A quarter of a century ago Emanuel de Kadt (1979: 19) pointed to the lack of attention paid to tourism development under state socialism. Response from the academic literature was slow in emerging, and only gained momentum as political change was taking place in the Soviet bloc (Allcock and Przecławski, 1990; EIU, 1991; Hall, 1991b). This was the result of logistical problems of empirical research, the relatively low priority given to, and poor documentation of tourism development in most state socialist societies, poorly developed conceptual frameworks for analysis, and the relatively small number of

academics holding an appropriate convergence of interests (Hall, 2001).

Yet there appeared to be a number of 'socialist' economic and political objectives which
tourism development was able to support.
These included generation of convertible currency, a redistribution of employment opportunities and promotion of positive place, regional
and national images and identity (Hall, 1984,
1991a). Tourism could also be viewed as an
instrument of foreign policy to be employed
explicitly as a propaganda tool (Qiao, 1995).
However, the very structural characteristics of
state socialism – centralized bureaucratic organization, inflexibility and antipathy towards indi-
vidualism and entrepreneurialism, resulting in
constraints on mobility and a low priority given
to service (industries), coupled with cold war
attitudes – including an alleged fear of Western
ideological contagion and social corruption,
were the very antithesis of the flexibility, respon-
siveness and market orientation required to
develop international tourism attractive to high-
spending markets. This suggested that a centralized 'command economy' and successful
international tourism development were logi-
cally incompatible. Indeed, this was reinforced
by the outstanding success of the only decentral-
ized 'socialist market economy', Yugoslavia,
which by the end of the 1980s was generating
international tourism receipts equivalent to the
total for the rest of CEE and the Soviet Union.

Domestic tourism and recreation was often
heavily subsidized particularly, and sometimes
exclusively, for urban industrial workers and
bureaucrats. Although a residual of such collective provision remained in the early years of
post-communism in some countries to provide
low-cost holidays (Williams and Baláž, 2001),
ability to pay became the major criterion in
access to domestic tourism exposed to the dic-
tates of adopted market economy templates.
Participation was now severely influenced by
national and local inflation, unemployment, privatization (in its various forms), subsidy with-
drawal, and the imposition of sometimes high
rates of VAT. New constraints modified pathways
of outbound tourism. The 'Iron Curtain' was
replaced by a 'dollar curtain', lowered by potential Western host countries out of fear of a flood
of Eastern migrants. The need to secure exit
visas from home countries was largely removed.
However, despite bilateral and EU-wide agree-

ments, the requirement for hard currency entry
visas, often costing more than a month's income,
was imposed by Western governments for citizens of several post-communist and all remaining state socialist countries. The process of
accession to the EU would help to change that.

While the English language literature on
tourism development within such post-communist processes has been substantial (see Hall,
1998, for an earlier review of the literature),
there have persisted criticisms that the conceptual and theoretical strength informing much of
this work has been less than adequate.

The 'T' word(s)

Within the comparative political economy literature, Hanson (1998) argued that the lack of
conceptual attention paid to accommodating
'transitions' from communism to post-communism resulted from an inability to accommodate
adequately the particular dynamics of formation
and decline of 'the Stalinist socio-economic system'. None of the three dominant frameworks of
political economy – modernization, world-systems (see Chapter 1) and rational choice – were
seen to be able to account sufficiently for the
diversity of post-communist economic and
social dynamism, although each could relate to
a limited range of post-communist experience:
modernization theory appearing to accommodate processes taking place in Central Europe,
world-systems theory illuminating forms of
'dependency' emerging in Central Asia and the
Caucasus, and rational choice analysis assisting
an understanding of the 'devolution' of state
structures in countries where the nomenklatura
– party, bureaucracy and managerial elites –
were able to maintain control, directly or indirectly, of most of the national wealth. Thus,
because of the several modes of 'communism'
and 'post-communism' which have existed, the
validity of concepts of communist and post-
communist landscapes (Rugg, 1971, 1985,
1994), heritages (Light, 2000a,b), and contexts
for tourism and recreation, can be contested.

Yet 'transition' itself is a problem.
Particularly in relation to processes within the
Soviet bloc since 1989, the 't' word has been
employed (not necessarily consciously) to represent a prescriptive set of uni-linear conceptual attitudes which locate the shift away from

central economic planning almost exclusively in terms of market-oriented reform. Many (Western) institutions have come to view 'transition' and 'transitional' in a particular way, as the focus of, and vehicle for, prescriptively defined pathways towards a capitalist global agenda. For example, the United Nations' interpretation of 'countries with economies in transition' specifically relates to the former state socialist societies of Central and Eastern Europe (CEE) and the former Soviet Union (FSU), despite the caveat that:

> The designations 'countries with economies in transition' and 'countries with established market economies' do not express a judgement about the political system prevailing in the concerned countries or areas.
>
> (United Nations, 2002: ii)

Strictly defined, 'transition' is concerned with moving between two known points, the final dimension of which in this case is the integration, or prospect of integration, of former communist states into the world economy, notably through EU accession (Agnew, 2000). Yet 'transition' is an holistic process, and the initial social, cultural and psychological pre-conditions of each region and country should be incorporated into the modelling process as well as political and economic structures. Often such elements have been ignored or marginalized in the transition economic literature (Marangos, 2003). Tomer (2002: 421) emphasizes other shortcomings:

> Transition has generally been conceived of as a substitution of the organizational structures and the legal, financial and political relationships of capitalism for those of socialism, a replacement of 'hard' features. This conception leaves out 'soft' factors such as attitudes, behavioural orientations, values and beliefs which, for successful socio-economic performance, must mesh with the hard elements. Sufficient attention should be paid to intangible capital formation that creates new soft features.

Pursuing this argument, Lesley Roberts evaluates the nature and role of such 'soft' elements within the context of 'social capital' in the next chapter.

As an alternative framework concept to 'transition', 'transformation', although embracing notions of fundamental structural change, is less concerned with an end state, being open-ended and allowing for the substantial (converging and diverging) differences which exist between former communist countries. As a negotiated approach to the unknown, with an emphasis on means rather than ends (Saltmarshe, 2000), 'transformation' implies more flexible, less dogmatic and certainly less prescriptively econocentric philosophical approaches to social change at a number of levels, which respect cultures, sovereignty and peoples' apprehensions, and which can be imbued with ideals of sustainability and equality that cannot easily be accommodated within 'transition'.

In what he refers to as 'post-communist systems', Intriligator (1998) argued that both democratic and non-democratic, and collapsing and expanding economies can be accommodated, as simplistically exemplified in Fig. 2.1. He specifically pointed to the apparent failure of the 'shock therapy' economic policies for transition to a market economy in Russia, which adopted the 'Washington consensus package' of stabilization, liberalization and privatization (SLP), as recommended by the World Bank and the IMF. By contrast he points to the success of Deng's economic policies for transition to a market economy in China being related to a rejection of SLP in favour of the institutions, competition and government (ICG) approach. Vietnam, whose political economy draws on South-east Asian, Chinese and Leninist cultural elements, has also experienced 'non-democratic' 'transition' to a market economy with no change in political regime occurring: what Fforde (2002) has referred to as a 'conservative' transition. This is implicitly close to notions of transformation, and does not assume an (interdependent) outcome of both economic and political change. According to Intriligator (1998), those nations in which both political and economic reforms have been simultaneously successful, such as the Czech Republic, are generally small, homogeneous European nations with: (a) an historic work ethic, (b) a relatively short period of socialism, and (c) support from neighbouring advanced industrialized nations. He argues that there is great difficulty in simultaneously establishing both democracy and a market economy without a very special set of circumstances being present.

Pathway analyses

One attempt to understand and account for the complexities of post-communist 'transition'

Fig. 2.1. Four possibilities for post-communist political and economic systems exemplified (source: Intriligator, 1998).

created both by diversity and by the persistence of some communist-period institutions, practices and mentalities, has been by means of the path dependence approach (Stark, 1992: Linz and Stepan, 1996; Stark and Bruszt, 1998). This allows conceptualization based on the premise that each country has a distinctive and unique path of extrication from state socialism: that particular variables play a crucial role in facilitating certain outcomes while constraining others (Meurs and Begg, 1998).

However, pathway analysis cannot assume that outcomes are determined or predictable since social and technological change continuously influence choices and decisions (Levi, 1996). By highlighting individual countries' and regions' explicit and implicit political, cultural and social continuities and discontinuities, comparative analyses of different paths question the emphasis upon rapid and wholesale marketization of economic life as the focus of post-communist restructuring. Indeed, examination of sub-national trajectories – impacting on specific cities, towns, villages and peripheral regions (e.g. Eikeland and Riabova, 2002) – illuminates local adaptations to fluid political, economic and cultural situations. This is important as 'transition' processes may actually reinforce and rejuvenate long-standing core–periphery, urban–rural, class, ethnic, gender and regional inequalities that may persist from situations of rural and regional exploitation under central planning (Staddon, 1999, 2001).

The way in which state power has been taken into private hands has been an important factor determining the nature of post-communist transformation, influencing, for example, the nature and extent of enterprise restructuring (Uhlenbruck et al., 2003). The concept of embeddedness may provide a useful analytical tool for understanding such processes, setting overtly economic activity within a framework whereby it can be seen to be shaped by social and political processes (Riley, 2000). The post-communist privatization of power has often been pervaded by vertical and horizontal networks of reciprocity – 'survival networks' (Kewell, 2002) – which have their roots in the communist period, and which may support or impede restructuring processes (Grabher and Stark, 1997), such as the privatization of hotels and other tourism assets and infrastructure (e.g. Koulov, 1996). In situations of persisting communism, such 'adaptive social relations' have been recognized in Vietnam (Fforde, 2002) as comparable to the 'competitive clientelism' observed in South-east Asia (e.g. Hutchcroft, 1997), and were seen to be preserved and augmented during transition. Case study examples include an examination of the transfer of Western management techniques to tourism enterprise managers in Russia (Phillips, 1996: 163). Two groups were recognized as 'gatekeepers' of structural transformation: (a) 'new' tourism managers who relied on old vertical structures and entrenched cultural expectations; and (b) those entrepreneurs who, while apparently comfortable with new conditions, were likely to be 'reconstructed' communist *nomenklatura*, as recognized, for example, by Harrison (1993) in Bulgaria. In some instances, particularly where the institutions of civil society have not been firmly established, such privatization of power can result in the development of mafia-type organizations (Anderson, 1995; Frisby, 1998), not least in tourism development and management.

Issues of 'conservative' transition

International tourism has only taken on a significant dimension in those post-Soviet continuing 'communist' societies which have modified their economic systems to accommodate demand-sensitive provision. China and Cuba are notable in this respect, although highly contrasting both in terms of scale and tourism history. Before the 1959 revolution Cuba acted as the Caribbean backyard playground of the USA, while pre-1949 China remained either relatively closed or in turmoil or both, and only from 1978 did the country's leadership begin to recognize the economic role of international tourism (Sofield and Li, 1998: 369) (see Chapter 12).

For both Vietnam and Cuba, continuing communism in the post-Soviet era has required readjustment from semi-colonial client state status (in relation to the Soviet Union) to an independent status exposed to the rigours of the (capitalist) global economy, of which international tourism is clearly an integral element. In Vietnam, of the four communist development stages recognised by Wu and Sun (1998) – independent socialism 1975–1977, orthodox socialism 1978–1985, glasnost socialism 1986–1990 and market socialism since 1991 – the 'orthodox' and glasnost phases were the direct result of Hanoi's dependency relationship with Moscow. Following the lifting of such hegemony, Wu and Sun (1998) argue, Vietnam's latitude for domestic and international policy making, including tourism development, increased significantly. This has been most explicitly expressed in a bilateral trade agreement with the USA which came into effect in December 2001, and which was anticipated to increase FDI between a third and a half by 2003 (Binh and Haughton, 2002).

Vietnam's cultural heritage, from ancient to colonial times, is substantial (e.g. Parenteau *et al.*, 1995), but the economic and social liberalization process *doi moi* has been heavily concentrated on the two main cities of Hanoi (Quang and Kammeier, 2002; Thomas, 2002) and Ho Chi Minh City (Drakakis-Smith and Dixon, 1997; Gainsborough, 2002). The spatial and structural inequalities in the country's development processes emerging from this, fuelled by spatially unequal flows of FDI, including those of tourism, have raised serious questions about the sustainability of national development policy (Mai, 2002) and have highlighted the urgent need to address a range of social and environmental problems (Bolay *et al.*, 1997). Luttrell's (2001) focus on livelihood diversification in coastal Vietnam, for example, has highlighted the socially and environmentally differentiated impacts that economic transition development processes bring even without political change.

Cuba was only drawn into the Soviet bloc from 1960; in the previous decades of the 20th century it had become a notorious playground for North Americans, for whom the island was easily accessible, with a winter climate widely considered to be healthy. Tourism became the country's third most important 'export' commodity after sugar and tobacco, and by the early 1940s Cuba was host to more than a third of all tourists visiting the Caribbean (Momsen, 1985). Visitor numbers reached their peak in 1957 when almost 356,000 foreign tourists and excursionists visited the country, 87% of whom were from the USA. This structural dependence was reflected in a spatial concentration of most tourism activity in and around the capital Havana and at the purpose-built north-coast resort of Varadero. This merely emphasized the dependency relationships – both internal and external – which characterized tourism's role, alongside sugar and tobacco. By reinforcing Cuba's semi-colonial status, it was such relationships that the revolution – with a little help from the subsequent US embargo – sought to eliminate (Hall, 1992: 109–111).

As a country substantially dependent upon a single primary product (sugar) and without major fuel and energy resources of its own, Cuba came to rely heavily upon the Soviet

Fig. 2.2. Contrasting recreational environments of Vietnam. Opposite: (a) The stunning offshore limestone landscapes of the southern coast. (b) In the Mekong Delta. (c) Queue for Ho Chi Minh's mausoleum, Hanoi. Above: (d) Exhibit in Hanoi's war museum.

Union as a ready market with a guaranteed price for much of its sugar crop and as a supplier of cut-price shipments of oil. With the break-up of the Soviet empire between 1989 and 1991, and Castro's outright rejection of *perestroika* as a guiding philosophy, Cuba needed to create new internal and external strategies in order to survive, economically and politically (Berrios, 1997). Economic decline here in the early 1990s was more dramatic than for the post-communist states undergoing 'shock therapy' in CEE, as output shrank to just 60% of 1989 levels (Pickel, 1998). As a consequence, the Cuban system of centralized economic planning and management was dismantled, and, refusing to adopt rapid shock therapy, the Cuban leadership employed variants of the Chinese and Vietnamese models, decentralizing economic activity while maintaining a tight rein on political structures.

Economic 'transition' experienced in Cuba between 1989 and 1995 (Everleny Perez, 1995) initially entailed crisis management, with economic adjustment to address major distortions. The 'special period in times of peace', 1989–1993, sought to reduce imports and increase efficiency in strategic sectors including tourism (Simon, 1995). From 1993, key reform measures were undertaken, including legislation encouraging certain forms of foreign investment, decriminalization of the use of hard currency and the authorization of self-employment in 150 private occupations. Many of these acts were clearly contrary to previous policies and principles (Berrios, 1997). Private rented accommodation and restaurants were legalized in 1995, and joint ventures with overseas interests (Spanish, Canadian, German, Mexican) grew to a greater extent in tourism than in any other sector. The emerging dual economy, however, has seen a widening gap between the traditional socialist sector using the Cuban peso and a dynamic internationalized sector, most notably tourism, operating in US dollars (Mesa-Lago, 1998). Further, anathema to Castro's puritan concept of a Cuban nation and a target for eradication by the revolution, sex tourism has returned (Davidson, 1996; Segre et al., 1997: 266). This ethically counter-revolutionary development results from the uneasy co-existence of an economic policy which places emphasis on international mass tourism growth while retaining low domestic incomes and restricted mobility, thereby inducing relative deprivation (Seaton, 1996, 1997) and demonstration effects. Tourist numbers and spending have increased dramatically, although while more than one and a half million tourists now visit Cuba annually (Table 1.2), such numbers could probably be trebled without the US embargo and consequent exclusion of the country from both the US market and the Caribbean cruise industry.

By contrast, international tourism policy in politically and economically unreformed North Korea appears to have shifted little from the centralized Stalinist model of severe prescription long pursued (Hall, 1986, 1990). Sea access from South Korea has been initiated, but plans for transport co-ordination between the two halves of the peninsula (Pak and Kim, 1996) have been slow. The White House

Fig. 2.3. Cuban heritage. (a) Iconic 1950s gas guzzler in Old Havana World Heritage Site. (b) Emphasizing infrastructural constraints: water delivery in Old Havana. (c) Limestone landscape of the Viñales valley tobacco-growing region.

belligerently pronouncing on an 'axis of evil' – of which DPR Korea is seen by Washington as a major component – has not assisted their progress (Ryan, 2002; Cha and Kang, 2003). The only minor example of injecting 'transitional' content into economic policy has been an attempt to emulate China's Shenzhen and other special economic zones by establishing a free zone in the border area with Russia and China. Despite a US$160 million casino, bank and hotel complex built by Hong Kong interests, at Rajin-Sonbong (Foster-Carter, 1998; Jo and Adler, 2002), this development has met with only limited success (Yeonchul, 2001; Koh, 2002).

Some analysts have argued that it could be possible for North Korea to retain a separate identity from the south, persisting with a collectivist political culture, but generating a variant of the Chinese, Vietnamese or Cuban economic transition model (e.g. Kimura, 1999). Certainly, continuing lack of economic openness – including highly prescriptive attitudes towards international tourism – has not assisted economic growth (Jin, 2003).

In China, as Chapter 12 suggests in the context of Hunan Province, tourism has emerged as a vehicle for mediating between some of the tensions of modernization resulting from the interplay of a rigid application of state socialism, the conservatism of tradition, and the reforming demands of economic development. By contributing to the modernization process, promoting heritage for product development, while meeting a number of socialist objectives, tourism has been seen to exert a centripetal influence in Chinese policy and society (Sofield and Li, 1998). However, by 2020 China is predicted to be the world's most important tourist destination with 137.1 million international arrivals, representing 8.6% of global market share and an annual growth rate from 1995–2020 of 8.0% (WTO, 1997). This compares with nearly 31.2 million arrivals in 2000 (Table 1.2) and suggests social and political impacts that can barely be foreseen. Further, China's domestic tourism has rapidly expanded in the 1990s, from 240 million domestic arrivals in 1989 to 640 million in 1996 (Wen, 1997), and Chinese tourism to Hong Kong increased considerably in the later 1980s and 1990s before that territory was reincorporated in 1997

(Zhang and Qu, 1996). Some of the local and regional pressures this brings, coupled with other domestic demographic dynamics, are represented in Chapter 13.

For Vietnam, Cuba and China, despite highly contrasting geographical and historical contexts, all three countries have managed to demonstrate that neither economic shock therapy nor multi-party political democracy are necessary ingredients for relatively successful economic change (Naughton, 1996). Their experience challenges Western prescriptive concepts of 'transition'. While not consistent in all three countries' transformations, the role of tourism has been both significant and distinctive.

The experience of South Africa has, of course, been different again. While abolition of apartheid was the explicit face of a transformation of political culture, the country's character of capitalist economic development and remaining deep divisions within society emphasize the significance of continuity alongside explicit change. Such a context for tourism development is clearly not unproblematic (Chapter 14).

Transition to EU Accession

As suggested in Chapter 1, the Euro-centric focus of post-communist 'transition' has functioned to assist and to act as a convenient precursor to the process of EU enlargement. Smith (2000) has suggested that the roles played by the EU in the negotiation of 'framework agreements' for the emerging European order expose contradictions in the EU's performance and raise questions about its capacity to shape the 'new Europe'. None the less, Dingsdale (1999: 149) has argued that we are witnessing a 're-territorialization of Europe', the key dynamics of which are:

- the consolidation of the European Union project as the dominant idea of Europe;
- EU territory conceived as the economic, social, political and cultural core of the continent; and
- a peripheral status for post-socialist Europe within it.

These factors could be seen to be articulated in:

- the terms defining which countries were fit to join the EU;
- their suppliant applications for membership;
- their compliance with EU rules for assistance; and
- their acceptance of disadvantageous terms of trade and the application of EU protocols.

And the rest of Europe? Conceptions of an 'integrating' west, a 'reforming' centre, and a 'struggling' east ignore north–south disparities such as those in the eastern periphery of the continent, exposing a relatively stable and co-operative north-east, contrasting with a fractious and still unstable south-east, despite the fact that

> The Aegean and Black seas are an alphabet soup of American and European projects, programmes, processes and partnerships, all designed to encourage good neighbourly relations and prepare the way for entry into Euro-Atlantic institutions.
>
> (King, 2001: 49)

Fig. 2.4. Tourism in North Korea. Opposite: (a) Prescriptive tourism: military lecture on Korea's continuing division at the Panmunjom de-militarized zone. (b) Twin-tower international hotel in Pyongyang. Above: (c) Mass rally and effective use of flash cards, Pyongyang stadium.

The Mediterranean dimension: tourism and political division

Triggered by a coup backed by military rulers in Athens designed to annex the island to Greece, Turkish military intervention in 1974 saw the invasion and occupation of 37% of Cyprus' territory – including some of the most popular tourist areas – and expulsion of 200,000 Greeks from that northern 'half' (Zetter, 1994). United Nations sanctions were subsequently imposed against this Turkish 'occupation'.

The division has persisted (Kliot and Mansfeld, 1997), with the Republic of Cyprus government representing the Greek Cypriot 'south', and the Turkish Republic of North Cyprus, declared in 1983 but recognized only by Turkey, acting on behalf of the 'north'. Turkish Cypriots want a confederation of Cyprus recognizing two states; Greek Cypriots want a reunified island with the return of all refugees. They certainly want unity of their country, but not necessarily of its people (Fisk, 1998), not least because some of the best (Greek Cypriot-managed) tourist resorts were in the north.

The island's partition focused attention on the role of tourism, which, in the 1970s and 1980s, took on a prime economic importance in the south (e.g. see Gilmor, 1989; Kammas and Salehiesfahani, 1992; Clements and Georgiou, 1998; Sharpley, 2001, 2003). With a UN economic boycott of Turkish Cyprus, however, precluding direct access except from Turkey, tourism in the north has been severely limited, receiving only around 10% of the island's total (Lockhart, 1994; Scott, 1995; Warner, 1999). Reasons for this are discussed in Chapter 10.

In July 1997 the European Commission in *Agenda 2000*, its communication to the European Parliament on the future development of the Union, confirmed that accession negotiations would begin, and reiterated the EU's determination to play a positive role in bringing about a settlement in Cyprus.

But the Cyprus problem has had the potential to confound negotiations for enlargement. Greece has repeatedly indicated that Athens would block expansion if Cyprus was excluded, arguing that the admission of other countries into the EU could not proceed if the reason for not admitting Cyprus was the island's political problem. The Greek position was that Cyprus could not be allowed to become a 'hostage' of Turkey, and that the EU must not accept a Turkish veto on the admission of Cyprus.

Further, until the Helsinki Summit of December 1999, Greece had vetoed any move towards inaugurating enlargement discussions with Turkey (Anon., 1998).

Since the end of the Cold War, Turkey's relationship with Europe has changed: Turkey's strategic importance to the West – particularly as a NATO listening post and frontline base – has declined, just when the EU has forged ahead with enlargement. Until the Helsinki Summit there had been growing tension between the EU and Turkey as the latter had watched CEE countries leapfrog its own application for membership.

For Turkish Cypriots there is no body that has a mandate to speak for both communities on the island, or to apply for EU membership and conduct accession negotiations with the EU on behalf of the whole of Cyprus. Further, agreeing to become a member of the EU, of which Turkey was not a full member while Greece was, would mean the destruction of the balance between the two motherlands over Cyprus, and tantamount to an indirect unification of the island with Greece.

As the Gordian knot of Cyprus' role within Europe begins to be loosened, it is clear that continued obstacles to Turkey's accession talks could lead to frustration in Ankara and a 'permanent' albeit not internationally recognized division of Cyprus. (Greek) Cyprus can prosper within the EU without any wider settlement of the island's political condition being instituted. Alternatively, if relatively rapid progress of domestic reforms within Turkey took place, and fast-track EU membership resulted, agreement on the re-establishment of a confederated Cyprus could come about, but now within the auspices of the EU and overseen by the three guarantors (Greece, Turkey, the UK) as EU member states.

Ironically, even in the globalized 21st century, an island as small as Cyprus has potentially held one of the major keys to EU enlargement, eastern Mediterranean stability, and CEE incorporation into mainstream Europe.

Selection and compliance

As the most important of those institutions, the procedure for gaining membership to the EU is long and complicated (European Commission, 1998a, 1998b; Mayhew, 1998; Phinnemore, 1999). A country must begin diplomatic relations with the EU, sign prescriptive trade and cooperation agreements, and develop a Europe Agreement. Once this Agreement comes into force, the country can formally apply for membership. This involves submitting a lengthy document for consideration by the European Commission which in its turn prepares an opinion on the applicant's characteristics in relation to accession criteria. These, agreed in 1993, at the European Council meeting in Denmark and referred to subsequently as the 'Copenhagen criteria', required that each candidate country should achieve:

- stability of institutions guaranteeing democracy, the rule of law, human rights and respect for and protection of minorities;
- the existence of a functioning market economy as well as the capacity to cope with competitive pressure and market forces within the EU;
- the ability to take on the obligations of membership including adherence to the aims of political, economic and monetary union; and
- the conditions for its integration through adjustment of its administrative structures, so that European Community/Union legislation transposed into national legislation is implemented effectively through appropriate administrative and judicial structures.

When most of the criteria have been met or seem likely to be met in the short term, formal accession negotiations begin. At the conclusion of these the country is invited to join. The last step in the process is to hold a referendum in the country on the question of joining the European Union. If a majority of the voting population affirm, the country can gain access.

During the period 1988–1997 ten CEE countries submitted applications to the EU: Hungary, Poland, Czechoslovakia (which divided in 1991 into the Czech Republic and Slovakia), Romania, Bulgaria, Estonia, Latvia, Lithuania and Slovenia; Cyprus, Malta and Turkey were also viable candidates whose applications had been submitted earlier. In 1997 the European Council decided to treat the candidate countries in two groups. The so-called 'first wave' or 'fast-track' of countries was to consist of the Czech Republic, Hungary, Poland, Slovenia and Estonia plus Cyprus, while the 'second wave' group included the other five CEECs plus Malta.

Spurred on by the Kosovo conflict of 1998/99 and its aftermath, which re-emphasized the need to stabilize and eventually embrace the states of south-eastern Europe, and to provide more effective and efficient means of pan-European coordination, EU president Romano Prodi called for a speeding up of the enlargement process. In December 1999 at the Helsinki European Council ('EU leaders' summit') meeting, Latvia, Lithuania, Slovakia, Bulgaria, Romania and Malta were elevated to fast-track status, while Turkey was rendered a candidate country (European Commission, 1999c).

This last decision marked a significant moderation in Greek opposition to Turkish membership, but the Greek government did not accept the notion of eventual Turkish membership without imposing conditions. Turkey would have to agree to take any territorial disputes between the two countries in the Aegean to the International Court of Justice. Most importantly, it also faced the prospect of Cyprus becoming an EU member without a political settlement for the divided island (see Chapter 10). Turkey would need to improve its human rights record and modernize sections of its economy before EU membership drew closer. Candidate status did not inaugurate accession negotiations, but permitted the country to benefit from the type of accession agreement which would generate funding in subsequent years.

The 'Europe Agreements' preparing for accession have provided the framework for bilateral relations between the EU and the partner countries, and cover trade-related issues, political dialogue, legal approximation and other areas of cooperation, including industry, environment, transport and customs, the free movement of services, payments and capital in respect of trade and investments, and the free movement of workers. They aimed progressively to establish a free-trade area between the EU and the associated countries over a given period.

Necessary assistance

The PHARE programme was identified as the financial instrument specifically aimed at helping achieve the objectives of the Europe Agreements. By providing funding to enable the partner countries to prepare for membership of the EU, PHARE has been fundamental to the process of integration and enlargement (European Commission, 1999b), funding projects emphasizing education, training and institution capacity-building.

Together with PHARE and aid for agricultural development, structural aid for the applicant countries for the period 2000–2006 would be directed towards aligning them on EC infrastructure standards, particularly in the transport and environmental spheres. This became known as ISPA (Instrument for Structural Policies for Pre-Accession), the responsibility of DG Regional Policy. Over half of the total ISPA budget is allocated to Poland and Romania, by far the two largest countries of the region. However, as a percentage of GDP, Poland has received less than a third as much as Latvia.

SAPARD (the Special Accession Programme for Agriculture and Rural Development) has aimed to help candidate countries deal with the problems of the structural adjustment in their agricultural sectors and rural areas, as well as in the implementation of the *acquis communautaire* concerning the CAP and related legislation (European Commission, 1999a). Thus with some anxiety already expressed about the size of the farming sector in CEE, SAPARD funds were not intended as production support subsidies, but to encourage rural diversification away from agriculture. This has embraced the upgrading of local infrastructure, support for education and training, assistance for the establishment of off-farm enterprise, and improvement in sales channels and quality controls for the food industry. Again, SAPARD was budgeted from 2000 until the end of 2006, and has been available to candidate countries only up to the point of their accession to the EU.

According to the terms of accession (European Commission, 2003a,b), temporary compensation is paid by the EU to the Czech Republic (€389 million), Cyprus (€300 million),

Malta (€166 million) and Slovenia (€131 million) for the period 2004–2006 to compensate for the loss of such pre-accession support. Additionally, all ten acceding countries are being provided with a special temporary cash-flow facility totalling €2.4 billion for the period 2004–2006 to assist their budgetary positions (Table 2.1). These amounts result in receipt of €1 billion for Poland and €100 million for the Czech Republic after corresponding reductions in the level of structural funds normally available to them (European Commission, 2003a: Chapter 29).

Significance for tourism

The necessary restructuring of EU programmes to accommodate eastern and Mediterranean enlargement will significantly influence policy impacts on tourism and recreation in the longer term. Tourism is recognized in Europe as making a valuable contribution to regional development and job creation, and initiatives receive support from a range of EU programmes, the major sources of which are indicated in Table 2.2. These represent both regional and local level scales of activity, and in relation to the LEADER+ programme, for example, the question has been posed as to the extent to which strategies based on a bottom-up philosophy can provide the necessary regional structures required to realize the regional potential implicit in the goals of integrated territorial development and of inter-territorial and transnational cooperation (Roberts and Hall, 2001: 74).

Williams and Baláž (2004) suggest that in relation to Slovakia's EU accession transition:

- the most significant institutional and allocative changes in the transformation of tourism had already occurred prior to accession to the EU;
- there has been more continuity than change in many aspects of both domestic and international tourism during the shift from state socialism to a market economy; and
- while continuity is expected to be the dominant feature of the post-accession trajectory of tourism development, it is possible to identify a number of significant potential changes: the mediation of cross-border flows

to new, existing and non-EU members, and sustained growth in tourism demand, contingent on economic convergence.

Such apparent continuity, rather than change, indicates the extent to which, in the 15 years since state socialism, the former Soviet bloc states of Central Europe (Czech Republic, Hungary, Poland, Slovak Republic) and Slovenia have become compliant to 'European norms'. In this respect they are set alongside Malta and (southern) Cyprus, and suggest that tourism represents an indicative benchmark. By contrast, patterns of tourism development in Bulgaria and Romania, 'fast track' candidates not acceding in 2004, do not present an image of continuity.

Regional cross-border mobility is a significant by-product of post-communist restructuring. It reflects forces of both integration – tourism, labour migration, shopping – and disintegration, or at least dislocation – refugee flight and the need for informal petty trading and exchange (Konstantinov et al., 1998). Indicative of an intricate web of linkages and networks on both sides of borders, such mobility experiences have expanded dramatically (Konstantinov, 1996), raising new questions for local and regional integration (Böröcz, 1996; Iglicka, 1998; Kratke, 1998; Hall, 2000; Williams and Baláž , 2000).

Each of the candidate countries varies widely with respect to size, population, economic performance (Table 1.1), political institutions, and environmental state (Saiko, 1998; Williams and Bjarnadottir, 1998). The 2004 enlargement process poses a unique challenge, since it is without precedent in terms of the scope and diversity of the number of candidates, their area (an addition of almost a third of EU territory), population (an increase of 75 million), and wealth of histories and cultures.

Defining and Redefining Identity Through Tourism

It should hardly require repeating that each 'transition' country has its own and often divergent cultural characteristics and aspirations, patterns of historical evolution, levels of economic and infrastructural development, spatial

Table 2.1. Special temporary cash-flow facility, 2004–2006, for EU acceding countries.

Country	Cyprus	Czech Republic	Estonia	Hungary	Latvia	Lithuania	Malta	Poland	Slovakia	Slovenia
Sum in €million at 1999 prices	38	358	22	211	26	47	66	1443	86	101

Source: European Community, 2003a: Chapter 29.

Table 2.2. EU funding sources for tourism.

Programmes	Purpose
Structural funds: ● European Regional Development Fund (ERDF) ● European Social Fund (ESF) ● European Agricultural Guidance and Guarantee Fund (EAGGF)	Major funds for promoting regional, economic and social development
LEADER+	To encourage local mutual support for the implementation of integrated, high-quality strategies for sustainable rural development, which include: ● enhancing natural and cultural heritage ● reinforcing economic environments ● community capacity building
Other miscellaneous	● Research and technology development ● Environmental protection ● Improving the competitiveness of SMEs ● Preservation of cultural heritage

structures, environmental attributes and political constructs. Such differentiation was illuminated in the early 1990s by the dismantling of federal structures – one state socialist (the Soviet Union), and two which had emerged from the aftermath of World War I (Czechoslovakia and Yugoslavia). Many of the new, often small, independent states arising out of the fragmentation were eager to establish and articulate individual national identity and statehood. International tourism and marketing imagery has been one vehicle through which such identity has sought expression (Hall, 1999, 2002).

The inhabitants of post-socialist Europe have found themselves in a vortex of contesting cultural influences (Dingsale, 1999: 146). Such influences are characteristic of all the societies highlighted in this volume, and it is clearly expressed across a wide range of cultures that in the domestic and international uncertainties of 'transformation', tourism's role may assist in a strengthening or re-affirmation of identity at a number of levels: national, ethnic, destination/place and gender/household.

This 'hierarchy', in its turn, is characterized and fragmented by differing conceptions of nation and of national collective identity, individual attitudes and value orientations, and constant change in the meaning and structure of 'identity' (Pickel, 1997: 203).

National imagery and identity

The identity-building requirements of the global marketplace emphasize the importance of three sets of interrelationships:

● that between imagery and the need to respond to new political and economic circumstances;

Fig. 2.5. Romania: petty trading on a major transit route through the Carpathians.

Fig. 2.6. Slovenia: local cross-border bus from Trieste, Italy.

- that between national identity and tourism branding; and
- that between economic and political re-imaging requirements.

The political dimension focuses on the need to differentiate a country clearly from its neighbours, a function with important economic dimensions for attracting tourism and inward investment, entailing:

- a 'Europeanness' for those countries acceding or seeking accession to the EU;
- disassociating from the recent past, but perhaps, ironically, drawing on heritage of a previous era; and

- disassociating from regional instability.

The economic dimension is concerned with the need to respond to new and changing market demands and increasing market differentiation, by:

- reassuring former markets that quality and value had been restored, even if the destination in question had just lived through a bitter conflict and had changed its political status and name in the process – as well exemplified in Croatia's 'A New Welcome. An Old Friend' promotion, portraying this 'new' country as a familiar and safe destination, comfortably European and an integral component of successful Mediterranean tourism, one of the mainstream icons of the European leisure experience (Hall, 1999);
- generating customer loyalty (through repeat visits), and raising *per capita* tourism income levels (through targeted high-quality product and service development); and thereby
- securing long-term competitive advantage through the country's major tourism attributes.

The conjunction of European economic and political convergence, and the leisure search for new experiences and products, provides a potentially wide range of contexts for the interweaving of national imagery and the promotion of tourism. This was well exemplified when, following its establishment in 1996, the Slovenia Tourist Board's mission statement firmly intertwined national political and tourism aspirations:

> To promote Slovenia as a country with a clear and distinctive identity and clearly defined comparative and competitive advantages and thereby assist the Slovene economy by marketing Slovene tourism in a concrete manner.
>
> (Slovenia Tourist Board, 1998: 1)

In their tourism marketing, most countries have adopted 'straplines' to encapsulate desired national brand images with underlying implications for European acceptability: uniqueness, accessibility, security and ecological friendliness (Table 2.3). However, the bland, cliché-ridden and even untenable nature of some of these may suggest that their adoption and the imagery it generates has sometimes been poorly advised.

New opportunities have been opened by technological development. Significantly, much effort has been put into producing lively and imaginative websites, with marketing messages and logos which have emerged as branding tools in their own right.

None the less, identity and image generation and reinforcement can raise contradictions and tension between the generation of collective identity for a country's citizens, to forge domestic unity and common aspirations – which may be based on ethnic or national particularism and even exclusivity – and an acceptable image for international tourist and inward investors' consumption. Tourism promotion of multi-cultural heritage as an essential ingredient of Serbia's identity might have had some difficulty being accepted by many Serbians during the dark days of the 1990s:

> Serbia is the meeting place of cultures, religions and languages. Although more than forty different nations live in Serbia, they do have some things in common – their homes are wide open to friends.
>
> (NTOS, 2000)

Further, a sense of identity may be forged by external as well as domestic factors. Thus Cuba's Latin American and Caribbean cultural identity has been unwittingly reinforced by the continuing uncompromising US embargo policy, which has encouraged the imagery of collective self-help and near autarky by imposing the additional constraint on Cuba of an inability to access investment funds from the World Bank and the IMF (Leogrande, 1997). However, reinforcing identity through kinship,

Table 2.3. Selected examples of national tourism 'straplines'.

CEE country	Strapline
Albania	*Land of sun and hospitality*
Croatia	*Small country for a great holiday*
Czech Republic	*In the heart of Europe*
Hungary	*The essence of Europe*
Malta	*The heart of the Mediterranean*
Montenegro	*Ecological state*
Romania	*Come as a tourist, leave as a friend*
Slovakia	*A small country with a big heart*
Slovenia	*The green piece of Europe*

Cuba receives well over US$500 million per annum in remittances from Cuban-Americans (Avella and Mills, 1996).

Independence from the UK in 1960 did not stimulate an overall concept of Cypriot nationhood: citizens continued to think of themselves as Greek or Turkish Cypriots. Here was a country which had statehood thrust upon it largely as a compromise among foreign powers who quarrelled over its future. Since the island's partition in 1974, mass tourism in the south has helped to generate an external image of 'Cyprus' and to support a self image of development and 'Europeanness'. By contrast, since 1974, northern Cyprus identity has been more closely aligned to that of Turkey (see Chapter 10), and as such embraces elements of the 'other'.

Place promotion/destination identity: rural tourism portraying the sublime

Notable promotion has been pursued for urban 'cultural' destinations, with imaging strategies emphasizing (potentially paradoxically) 'European' heritage and progress, to appeal both to the tourism and wider economic investment markets (e.g. Goluza, 1996). Place promotion focusing on specific destination cities (e.g. Young and Kaczmarek, 1999), resorts or regions for tourism, wider investment and residential development purposes may be sensible for an industry which is seeking to promote niche products and activities, but it can easily convey a confusing message if lacking consonance with projected national imagery.

The current cultural tourism development emphasis on major urban centres is not sustainable, whether in the Czech Republic, Malta or Vietnam. The societies discussed in this volume contain many and diverse rural cultures which present myriad development opportunities for small-scale, high-income, locally controlled tourism generation.

Part of the escape strategy from the past and its images has been to diversify away from mass coastal tourism and to emphasize the uniqueness of cultural and natural resources (Bralić, 1995; Hall, 1999; Meler and Ruzic, 1999). Many rural areas are being re-imaged (Roberts

and Hall, 2001) in the promotion of traditional, 'idyllic' portrayals of timeless sustainability.

The negative impacts of political and economic change have often been most significant in rural areas, but with a spatially and structurally dynamic mix of mass and specialist markets to target, there are opportunities for rural tourism to act as a vehicle for rural identity. In 1995 the Romanian Ministry of Tourism recognized this and identified rural tourism as a major growth area (Light and Andone, 1996).

The re-imaging of rural areas, arising out of a commodification of the countryside and restructuring processes (Gannon, 1994), has witnessed growth in the past decade of eco- and nature tourism. For example, Poland moved to promoting itself as a 'natural' destination, with tourism literature becoming notably special interest-oriented (e.g. Witak and Lewandowska, 1996). Some segmentation, such as gastronomy (e.g. Csapo et al., 2000; Kraus, 2000; National Authority for Tourism, 2000), has been based on local economic back-linkages, particularly with restructuring rural economies. Other segments, such as health and spa resorts (e.g. Popesku, 2000) and heritage trails (e.g. Kornecki, 2000), draw on a fund of local cultural tradition which can be traced to classical times. By contrast, the rapidly growing activity holiday segments (e.g. Wieczorek, 1999; Stifanic, 2000) represent direct responses to perceived demand, although appropriate infrastructure and quality levels may not yet be in place.

'Heritage' occupies a more ambivalent branding role. Where the past is often drawn on to justify the present, 'heritage' tourism offers the irony of employing the past as an element of restructuring for the future, particularly for newly independent states drawing on their pre-communist roots. 'Heritage' is far from being a value-free concept: economic power and politics influence what is preserved and how it is interpreted (Lowenthal, 1997).

Promotion of rural and urban heritage, as an integral element of cultural history, was characteristic of the communist period (and earlier). Under the communists, however, such heritage promotion as the development of open-air 'village museums' (e.g. Focşa, 1970) tended not to be primarily for international tourism purposes but was meant to inculcate a sense of identity and pride amongst each country's own citizens.

The legacy of communism itself has become an aspect of heritage and has attracted international tourism interest. However, as Light (2000a,b) argues, this is a heritage which is defined and constructed largely externally. Within the destination countries themselves, he argues, there is usually little desire to remember the period of communism. Indeed, the legacy of this period is seen to be strongly dissonant with post-communist aspirations. This is expressed in the apparent lack of interest to interpret the legacy of communism for tourists, and in attempts to deny or remove the period from each country's past and to create a new imagery which is consistent with post-communist identity.

In the case of Slovenia, for example, needing both to establish an individual national identity and a suitable vehicle for (re-)attracting both established and new Western markets, in the early 1990s the country initially adopted 'The sunny side of the Alps' as a national tourism promotion 'strapline', embodying positive attractions of climate, topography and contiguity with Western Europe. Following Italian objections, this was replaced in the mid-1990s with 'The green piece of Europe', symbolizing a philosophical shift in terms of emphasizing, or at least suggesting, a small yet environmentally aware corner of Europe.

In both cases the country has tried to reinforce the imagery of Slovenia being firmly part of ('Western'/'civilized') Europe, and thus a consonance with EU accession, by emphasizing:

- a fashionable/politically correct ecologically friendly ethos;
- its Central European character: Hapsburg heritage, alpine associations, contiguity with EU members Austria and Italy; and
- disengagement from any 'Balkan' connotations.

For example, marketing hyperbole in a mass tourism brochure talks of Slovenia's 'delightful villages and warm and hospitable people, whose lives are still steeped in the traditions of centuries of Austrian rule' (Transun, 1998: 70). This suggests a Central/Western European heritage, untainted by more recent years. Just to emphasize the point, the

Slovenia Tourist Board bluntly states: 'Slovenia is actually situated in Central Europe not in Eastern Europe at all!' (Slovenia Tourist Board, 1999: 1). Yet the core objective of producing a consistent identity is constrained by at least three major factors:

- a lack of adequate finance to support appropriate marketing campaigns – a common trait – has been exacerbated by limited experience of global markets and a lack of appropriate expertise: pre-1989 mass markets were drawn from a relatively small number of West European countries;
- second, in a post-conflict context where rapid reconstruction and re-imaging is sought, tourism destination marketers may be pressured to return short-term results when long-term investment is required to build a consistent brand. Such a dilemma may be reflected in message inconsistency;
- third, while centralized, relatively authoritarian regimes can impose some control and coherence over the component parts of a destination image, it is more difficult within market economies to develop a coherent brand for destinations that are composed of myriad products and environments.

Gender/household identity: the context for rural women

The degree to which restructuring processes arising from transformation adversely or positively affect the role and identity of women is a matter of some debate. Weiner (1997) pointed to the effects of such processes for marginalizing and/or excluding women through reductions in state-funded health and child care, family planning and education, as well as heavy reductions in the administrative employment positions that women can find. Somewhat diffuse research has suggested generally that women are less well positioned to take advantage of new economic opportunities under such conditions. From the former East Germany for example, 80% of commuters to the West are male, a situation interpreted as reflecting women's limited ability to travel long distances to work because of family obligations (Rudolph et al., 1994).

The issue as to whether a woman's right to work, as guaranteed under communism, was a form of emancipation or simply one more aspect of state exploitation, has been much debated. Women often felt forced to work outside of the home, in addition to the inevitable domestic chores, and were likely to be treated at the workplace within the confines of sexual stereotypes. Yet, in the early years of post-communist restructuring in CEE, any economic and social advantages that women had gained under the communists were being eroded (Einhorn, 1991, 1993). For example, the probability of obtaining new employment in CEE has been calculated at three times more likely for a man than for a woman (Fong and Paull, 1993). In tourism, acting in such roles as guides and hostesses in newly privatized or new companies often means being paid a poor basic wage and having to depend on tips and other favours to sustain an adequate income. In tourism-related production such as crafts and folk arts a marginalization of women has also been identified (Buckley, 1997). Within the Maltese souvenir handicraft industry, Robson (2002) found identity to be promoted by the gendered images of local handicraft workers making textiles or forging metal and glass, and by gendered practices in the production, promotion, depiction and retailing of tourism crafts. Stereotyped ideas of authenticity reinforce patterns of gender inequality. Men tend to be wage earners, working in public occupations, while women's work is often informal and based in the home. Women dominate low-technology, nimble-finger craft production such as lacemaking and knitting. Geographically they are atomized and scattered. By contrast, men dominate the crafts employing technology, such as glass-blowing and metalworking, and are located in formal workplaces, which are, in turn, more accessible to the tourist gaze.

For entrepreneurial women, one response is to start their own business, despite the difficulties of obtaining credit and the risks of trying to survive alongside the grey economy, particularly when they often do not have the appropriate connections and, in post-communist societies, access to 'survival networks'. However, some have been supported by aid-assisted business incubators – facilities which can provide low-rent office space, access to information and communications technology and provide business training courses and advice. In Russia, women have received nearly half of the loans made to small businesses by a $300 million fund of the European Bank for Reconstruction and Development. In Hungary about 40% of new business in the later 1990s was started by women (Medvedev, 1998).

Tourism-related employment opportunities for local women are often typified by a predominance of unskilled, low-paid jobs. The gendered horizontal segregation of occupations is particularly noticeable in such semi-skilled, domestic and service-type occupations (Kinnaird and Hall, 1994, 1996, 2000; Swain, 1995), especially where they mirror functions carried out in the home. This is reinforced by assumptions that, if the best jobs and highest rewards are linked to an accumulation of human capital, women are inevitably disadvantaged because their process of accumulation is interrupted by marriage, birth and child-rearing (Coppock et al., 1995). Within the hotel and catering industry, for example, women tend to be offered work as unskilled, low-paid counter and kitchen staff, domestics and cleaners, while men are porters and stewards, but with a high proportion in professional, managerial or supervisory positions. None the less, Ghodsee (2003) argues that in the case of Bulgaria at least, state support for viable economic sectors such as tourism, with relatively high wages and employing a 'critical mass' of women, can help to ease women's burden within transition processes.

The intersection of gender and ethnic differentiation raises further issues of identity. In pre-Yugoslav war Montenegro, Devedzic (2002) saw tourism development transforming gender relations and roles to varying degrees in two distinct patriarchal host societies, Slavic Montenegrins and Muslim ethnic Albanians. Both groups were drawn into tourism, but the Montenegrin women found for themselves a greater sense of independence and identity than did the Albanian women, to some extent distancing themselves from the notion of tourism employment being an extension of feminized domestic work. In northern (Turkish) Cyprus, Scott (1995, 1997) found that although local (Muslim) women's participation in the tourism labour force had increased in recent years, migrant, and notably Romanian women, were being employed in those occupations considered 'unsuitable' by locals (in this case as croupiers in the casino). This division of female labour again was seen to highlight women's dual role and identity.

In the rural sector, tourism development can provide women with employment opportunities and support for self-determination. This can take the form of managing the provision of bed and breakfast accommodation (although this is often viewed pejoratively as simply an extension of the domestic role), management of farm-based attractions, or establishing one-person, family or collaborative SMEs (Petrin, 1996; Kulcsar and Verbole, 1997). Women often comprise the majority of rural populations, and tend to represent the oldest age groups, but their identity and status can vary enormously. In Slovenia two distinct groups of farm women have been identified: 'independent' and 'traditional' (Verbole and Mele-Petric, 1996). The former enjoy equal status with, or slightly privileged status compared to men. They are employed at home and have their own regular or occasional income; they make their own decisions and organize their own work such as agrotourism. 'Traditional' farm women, by contrast, do not enjoy equal status with their male relatives in the household or on the farm, but are usually restricted to making decisions that concern the household and children, and are limited or even restrained from decision making regarding the issues related to the farm and any pluriactivity, such as tourism.

Problems for rural women created by economic and political change have been related to structural levers on independence and autonomy (Siiskonen, 1996). Giving up the state and cooperative farms of communist times led to the loss of social services such as kindergarten as well as health and maternity care provided by the state and cooperative farms. Cultural activities similarly connected with these farms have also been lost.

Small-scale 'community'-based 'sustainable' rural tourism projects have been a notable element of European-aided projects in a number of post-communist countries. In the case of Albania, for example, women have been targeted, through the employment of participatory rural appraisal (PRA) in collaboration with

local NGOs. Schemes to reduce the danger of villages becoming impoverished in the backwash of large-scale coastal tourism development have sought to enhance local economic and social capacity and encourage financial institutions to support rural 'self-help' schemes (Fisher, 1996). However, community involvement and ownership of tourism development has often been inhibited by the legacy of minimal local experience of bottom-up development and lack of opportunities to participate in local decision making. This has been superimposed upon a tradition of male dominance, thereby doubly subjugating women in decision-making processes. Further, tensions may arise as the result of misunderstanding and incompatability of objectives. The promoters of 'ecotourism' projects, for example, may wish to present the absence of infrastructure as providing a pristine environment appealing to rural and ecotourists. Yet for local people, and especially women, participation in such schemes may be pursued on the assumption that tourism development will help to improve their local infrastructures and services, such as a clean water supply, piped sewage disposal facilities, electricity, an access road and telecommunications, and thereby reduce their daily burden (Holland, 2000).

In a number of transforming societies, several measures are required to ameliorate women's identity, role and employment status, if their human capital and self-fulfilment potential are to be realized. In South Africa, for example, women's rights were written into the new democratic constitution of 1996, yet the legacies of apartheid, including gender relations, have negatively influenced the post-apartheid pathway (e.g. Schäfer, 2001).

Required measures embrace women both as producers and as potential consumers of tourism (Weiner, 1997: 497). They include improved collection of data and information pertaining to women, attention by employment service providers to the specific needs of women, and specialized training and access to credit in order to encourage women's entrepreneurial activity. A basic need, particularly in less developed countries, is women's unfettered access to land and property (Bob and Musyoki, 2002; Hartl, 2003). The pivotal role of women in tourism, and of tourism in this process, places a considerable responsibility on the industry's development and management, and its collaboration and partnership with other social and economic development sectors.

Conclusion

Following on from the broader context setting of Chapter 1, this chapter has outlined and discussed three key themes. These are intended to provide frameworks for the following chapters' evaluation and exemplification of the nature and roles of, and influences on, tourism development in contemporary processes of 'transition'. The key themes were:

- the conceptual value of 'transition';
- the processes of 'transition' towards EU accession; and

- the definition/redefinition of identity through tourism.

In establishing these themes, this chapter has had two major objectives:

- to raise salient issues concerning the conceptualization of tourism within transition; and
- to bring analytical coherence to an understanding of the change processes being experienced in the range of 'transitional' and 'transformational' societies featured in this volume.

The following case study chapters (3–14) evaluate particular contexts for the intersection of tourism and transformation. The concluding chapter (15) will attempt to draw the book's major strands together and to revisit the value of a model of tourism transition.

References

Agnew, J. (2000) How many Europes? In: Hall, D. and Danta, D. (eds) *Europe Goes East: EU Enlargement, Diversity and Uncertainty.* The Stationery Office, London, pp. 45–54.

Allcock, J.B. and Przecławski, K. (1990) Special issue: Tourism in centrally-planned economies. *Annals of Tourism Research* 17(1).

Anderson, A. (1995) Organised crime, mafia and governments. In: Fiorentini, G. and Peltzman, S. (eds) *The Economics of Organised Crime.* Cambridge University Press, Cambridge, pp. 33–60.

Anon. (1998) Cyprus problem could threaten EU expansion. *Cyprus Mail*, 11 November.

Avella, A.E. and Mills, A.S. (1996) Tourism in Cuba in the 1990s: back to the future? *Tourism Management* 17(1), 55–60.

Berrios, R. (1997) Cuba's economic restructuring, 1990–1995. *Communist Economies & Economic Transformation* 9(1), 117–130.

Binh, N.N. and Haughton, J. (2002) Trade liberalization and foreign direct investment in Vietnam. *ASEAN Economic Bulletin* 19(3), 302–318.

Bob, U. and Musyoki, A. (2002) Gender geography in South Africa: facing challenges and exploring opportunities. *South African Geographical Journal* 84(1), 98–106.

Bolay, J.C., Cartoux, S., Cunha, A., Du, T.T.N. and Bassand, M.T.I. (1997) Sustainable development and urban growth: precarious habitat and water management in Ho Chi Minh City, Vietnam. *Habitat International* 21(2), 185–197.

Böröcz, J. (1996) *Leisure Migration.* Pergamon, Oxford.

Bralić, I. (1995) *The National Parks of Croatia,* 2nd edn. Skolska knjiga, Zagreb.

Buckley, M. (ed.) (1997) *Post-Soviet Women: From the Baltic to Central Asia.* Cambridge University Press, Cambridge.

Cha, V.D. and Kang, D.C. (2003) The Korea crisis. *Foreign Policy* 136, 20–29.

Clements, M.A. and Georgiou, A. (1998) The impact of political instability on a fragile tourism product. *Tourism Management* 19(3), 283–288.

Coppock, V., Haydon, D. and Richter, I. (1995) *The Illusions of 'Post-Feminism': New Women, Old Myths.* Taylor and Francis, London.

Csapo, K., Nagyvathy, E. and Pakozdi, J. (2000) *Hungary: Flavours of a Country.* VIVA Media Holding, Budapest.

Davidson, J.O. (1996) Sex tourism in Cuba. *Race and Class* 38(1), 39–48.

de Kadt, E. (1979) *Tourism: Passport to Development*. Oxford University Press, Oxford.

Devedzic, M. (2002) Ethnic heterogeneity and gender in the Yugoslav seaside tourist region. In: Swain, M.B. and Momsen, J.H. (eds) *Gender/Tourism/Fun(?)*. Cognizant Communication Corporation, New York, pp. 143–153.

Dingsdale, A. (1999) New geographies of post-socialist Europe. *Geographical Journal* 165(2), 145–153.

Drakakis-Smith, D. and Dixon, C. (1997) Sustainable urbanization in Vietnam. *Geoforum* 28(1), 21–38.

Eikeland, S. and Riabova, L. (2002) Transition in a cold climate: management regimes and rural marginalisation in Northwest Russia. *Sociologia Ruralis* 42(3), 250–266.

Einhorn, B. (1991) Where have all the women gone? Women and the women's movement in East Central Europe. *Feminist Review* 39, 16–37.

Einhorn, B. (1993) *Cinderella Goes to Market: Citizenship, Gender and Women's Movements in East Central Europe*. Verso, New York.

EIU (Economist Intelligence Unit) (1991) *Tourism in Eastern Europe*. EIU, London.

European Commission, The (1998a) *The European Union's Pre-Accession Strategy for the Associated Countries of Central Europe*. European Commission, Brussels.

European Commission, The (1998b) *The Challenge of Enlargement – the European Commission's Opinion, July 1997*. European Commission, Brussels.

European Commission, The (1999a) *Enlargement: Pre-Accession Strategy. Pre-Accession Assistance*. European Commission, Brussels <http://www.europa.eu.int/comm enlargement/pas/envir_transp.htm>.

European Commission, The (1999b) *Enlargement: Pre-Accession Strategy. The Europe Agreements*. European Commission, Brussels <http://www.europa.eu.int/comm/ enlargement/pas/europe_agr.htm>.

European Commission, The (1999c) *Helsinki European Council Presidency Conclusions*. European Commission, Brussels <http://www.europa.eu.int/council/off/conclu/dec99/ dec99_en.htm>.

European Commission, The (2003a) *Report on the Results of the Negotiations on the Accession of Cyprus, Malta, Hungary, Poland, the Slovak Republic, Latvia, Estonia, Lithuania, the Czech Republic and Slovenia to the European Union*. European Commission, Brussels <http://www.europa.eu.int/ enlargement/negotiations/pdf/ negotiations_report_to_ep.pdf>.

European Commission, The (2003b) *The Treaty of Accession 2003*. European Commission, Brussels <http://europa.eu.int/comm/enlargement/negotiations/treaty_of_accession_2003/table_of_content_en. htm>.

Everleny Perez, O. (1995) *Las Reformas Economicas en Cuba: un Analisis Critico*. Centro de Estudios de la Economia Cubana, Havana.

Fforde, A. (2002) Resourcing conservative transition in Vietnam: rent switching and resource appropriation. *Post-Communist Economies* 14(2), 203–226.

Fisher, D. (1996) Sustainable tourism in southern Albania. *Albanian Life* 59, 27–29.

Fisk, R. (1998) Divided they stand, separated by eternal hatred and suspicion. *The Independent*, 10 December, p. 10.

Focşa, G. (1970) *Muzeul Satului din Bucureşti*. Meridiane, Bucharest.

Fong, M. and Paull, G. (1993) Women's economic status in the restructuring of Eastern Europe. In: Moghadam, V.M. (ed.) *Democratic Reform and the Position of Women in Transitional Economies*. Clarendon Press, Oxford, pp. 217–247.

Foster-Carter, A. (1998) Asia's deadly enigma. *The Times Higher*, 4 September.

Frisby, T. (1998) The rise of organised crime in Russia: its roots and social significance. *Europe-Asia Studies* 50(1), 27–49.

Gainsborough, M. (2002) Understanding communist transition: property rights in Ho Chi Minh City in the late 1990s. *Post-Communist Economies* 14(2), 227–244.

Gannon, A. (1994) Rural tourism as a factor in rural community economic development for economies in transition. *Journal of Sustainable Tourism* 2(1-2), 51–60.

Ghodsee, K. (2003) State support in the market: women and tourism employment in post-socialist Bulgaria. *International Journal of Politics, Culture and Society* 16(3), 465–482.

Gilmor, D.A. (1989) Recent tourism development in Cyprus. *Geography* 74(3), 262–265.

Goluza, M. (1996) *Zagreb: the New European Metropolis*. Tourist Association of the City of Zagreb, Zagreb.

Grabher, G. and Stark, D. (eds) (1997) *Restructuring Networks in Post-Socialism: Legacies, Linkages and Localities*. Oxford University Press, Oxford.

Hall, D. (1984) Foreign tourism under socialism: the Albanian 'Stalinist' model. *Annals of Tourism Research* 11(4), 539–555.

Hall, D. (1986) A last resort: North Korea opens to tourism. *Inside Asia* 9, 21–23.

Hall, D. (1990) Stalinism and tourism: a study of Albania and North Korea. *Annals of Tourism Research* 17(1), 36–54.

Hall, D. (1991a) Introduction. In: Hall, D. (ed.) *Tourism and Economic Development in Eastern Europe and the Soviet Union.* Belhaven Press, London, pp. 3–28.

Hall, D. (ed.) (1991b) *Tourism and Economic Development in Eastern Europe and the Soviet Union.* Belhaven Press, London.

Hall, D. (1992) Tourism development in Cuba. In: Harrison, D. (ed.) *Tourism and the Less Developed Countries.* Belhaven Press, London, pp. 102–120.

Hall, D. (1998) Central and Eastern Europe. In: Williams, A.M. and Shaw, G. (eds) *Tourism and Economic Development in Europe.* John Wiley & Sons, Chichester, UK, pp. 345–373.

Hall, D. (1999) Destination branding, niche marketing and national image projection in Central and Eastern Europe. *Journal of Vacation Marketing* 5(3), 227–237.

Hall, D. (2000) Cross-border movement and the dynamics of 'transition' processes in South-eastern Europe. *GeoJournal* 50(2–3), 249–253.

Hall, D. (2001) Tourism and development in communist and post-communist societies. In: Harrison, D. (ed.) *Tourism and the Less Developed World: Issues and Case Studies.* CAB International, Wallingford, pp. 91–107.

Hall, D. (2002) Branding and national identity: the case of Central and Eastern Europe. In: Morgan, N., Pritchard, A. and Pride, R. (eds) *Destination Branding – Creating the Unique Destination Proposition.* Butterworth-Heinemann, Oxford, pp. 87–105.

Hanson, S.E. (1998) Development, dependency, and devolution: the anomalous political economy of communist and postcommunist societies. *Environment and Planning C: Government and Policy* 16(2), 225–246.

Harrison, D. (1993) Bulgarian tourism: a state of uncertainty. *Annals of Tourism Research* 20, 519–534.

Hartl, M. (2003) *Rural Women's Access to Land and Property in Selected Countries.* Sustainable Development Department, Food and Agriculture Organization of the United Nations, Rome <http://www.fao.org/sd/2003/PE07033a_en.htm>.

Holland, J. (2000) Consensus and conflict: the socioeconomic challenge facing sustainable tourism development in southern Albania. *Journal of Sustainable Tourism* 8(6), 510–524.

Hutchcroft, P.D. (1997) The politics of privilege: assessing the impacts of rents, corruption, and clientelism on Third World Development. *Political Studies* 45, 639–658.

Iglicka, K. (1998) The economics of petty trade on the Eastern Polish border. In: Iglicka, K. and Sword, K. (eds) *Stemming the Flood: the Challenges of East–West Migration for Poland.* Macmillan, London.

Intriligator, M.D. (1998) Democracy in reforming collapsed communist economies: blessing or curse? *Contemporary Economic Policy* 16(2), 241–246.

Jin, J.C. (2003) Openness and growth in North Korea: evidence from time-series data. *Review of International Economics* 11(1), 18–27.

Jo, J.C. and Adler, S. (2002) North Korean planning: urban changes and regional balance. *Cities* 19(3), 205–215.

Kammas, M. and Salehiesfahani, H. (1992) Tourism and export-led growth – the case of Cyprus, 1976–1988. *Journal of Developing Areas* 26(4), 489–506.

Kewell, B. (2002) Hidden drivers of organisational transformation in Poland: survival networks amongst state owned and privatized firms in the early 1990s. *Journal for East European Management Studies* 7(4), 373–392.

Kimura, M. (1999) From fascism to communism: continuity and development of collectivist economic policy in North Korea. *Economic History Review* 52(1), 69–87.

King, C. (2001) The new Near East. *Survival* 43(2), 49–68.

Kinnaird, V. and Hall, D. (eds) (1994) *Tourism: a Gender Analysis.* John Wiley & Sons, Chichester, UK.

Kinnaird, V. and Hall, D. (1996) Understanding tourism processes: a gender-aware framework. *Tourism Management* 19(2), 95–102.

Kinnaird, V. and Hall, D. (2000) Theorizing gender in tourism research. *Tourism Recreation Research* 25(1), 71–84.

Kliot, N. and Mansfeld, Y. (1997) The political landscape of partition: the case of Cyprus. *Political Geography* 16(6), 495–521.

Koh, H. (2002) Prospects for opening in North Korea and inter-Korean economic cooperation. *Asian Perspective* 26(3), 91–109.

Konstantinov, Y. (1996) Patterns of reinterpretation: trader-tourism in the Balkans (Bulgaria) as a picaresque metaphorical enactment of post-totalitarianism. *American Ethnologist* 23(4), 762–782.

Konstantinov, Y., Kressel, G.M. and Theun, T. (1998) Outclassed by former outcasts: petty trading in Varna. *American Ethnologist* 25(4), 729–745.

Kornecki, M. (2000) *The Trail of Wooden Architecture.* Krakówska Agencja Rozwoju Turystyki S.A., Kraków.

Koulov, B. (1996) Market reforms and environmental protection in the Bulgarian tourism industry. In: Hall, D. and Danta, D. (eds) *Reconstructing the Balkans.* John Wiley & Sons, Chichester, UK, pp. 187–196.

Kratke, S. (1998) Problems of cross-border regional integration: the case of the German-Polish border area. *European Urban and Regional Studies* 5(3), 248–262.

Kraus, V. (2000) *South Moravian Vineyards and Wine Cellars.* Vinarskou Akademii Valtice, Valtice.

Kulcsar, L. and Verbole, A. (1997) *National Action Plans for the Integration of Rural Women in Development: Case Studies in Hungary and Slovenia.* Food and Agriculture Organization of the United Nations, Rome.

Leogrande, W.M. (1997) Enemies evermore: US policy towards Cuba after Helms-Burton. *Journal of Latin American Studies* 29, 211–221.

Levi, M. (1996) Social and unsocial capital: a review essay of Robert Putnam's 'Making Democracy Work'. *Politics and Society* 24(1), 45–55.

Light, D. (2000a) An unwanted past: contemporary tourism and the heritage of communism in Romania. *International Journal of Heritage Studies* 6(2), 145–160.

Light, D. (2000b) Gazing on communism: heritage tourism and post-communist identities in Germany, Hungary and Romania. *Tourism Geographies* 2(2), 157–176.

Light, D. and Andone, D. (1996) The changing geography of Romanian tourism. *Geography* 81(3), 193–203.

Linz, J. and Stepan, A. (1996) *Problems of Democratic Transition and Consolidation.* Johns Hopkins University Press, Baltimore, Maryland.

Lockhart, D. (1994) Tourism in Northern Cyprus – patterns, policies and prospects. *Tourism Management* 15(5), 370–379.

Lowenthal, D. (1997) *The Heritage Crusade.* Viking, London.

Luttrell, C. (2001) Institutional change and natural resource use in coastal Vietnam. *GeoJournal* 55(2-4), 529–540.

Mai, P.H. (2002) Regional economic development and foreign direct investment flows in Vietnam, 1988–98. *Journal of the Asia Pacific Economy* 7(2), 182–202.

Marangos, J. (2003) Global transition strategies in Eastern Europe: moving to market relations. *Development* 46(1), 112–117.

Mayhew, A. (1998) *Recreating Europe: the European Union's Policy Towards Central and Eastern Europe.* Cambridge University Press, Cambridge.

Medvedev, K. (1998) A review of women's emancipation in Hungary: limited successes offer some hope. *Transition,* 26 February.

Meler, M. and Ruzic, D. (1999) Marketing identity of the tourist product of the Republic of Croatia. *Tourism Management* 20, 635–643.

Mesa-Lago, C. (1998) Assessing economic and social performance in the Cuban transition of the 1990s. *World Development* 26(5), 857–876.

Meurs, M. and Begg, R. (1998) Path dependence in Bulgarian agriculture. In: Pickles, J. and Smith, A. (eds) *Theorising Transition.* Routledge, London, pp. 243–261.

Momsen, J. (1985) Tourism and development in the Caribbean. *Mainzer Geographische Studien* 26, 25–36.

National Authority for Tourism (2000) *Romanian Gastronomy.* National Authority for Tourism, Bucharest.

Naughton, B. (1996) The distinctive features of economic reform in Vietnam and China. In: McMillan, J. and Naughton, B. (eds) *Reforming Asian Socialism: the Growth of Market Institutions.* The University of Michigan Press, Ann Arbor, Michigan, pp. 273–296.

NTOS (National Tourism Organization of Serbia) (2000) *Landscape Painted From the Heart.* NTOS, Belgrade <http://www.Serbia-info.com/ntos/inf_gen.htm>.

Pak, M.S. and Kim, T.Y. (1996) A plan for co-operation in transport between South and North Korea. *Transport Reviews* 16(3), 225–241.

Parenteau, R., Charbonneau, F., Toan, P.K., Dang, N.B., Hung, T., Nguyen, H.M., Hang, V.T., Hung, H.N., Binh, Q.A.T. and Hanh, N.H. (1995) Impact of restoration in Hanoi French colonial quarter. *Cities* 12(3), 163–173.

Petrin, T. (1996) *Basic Facts on Rural Women in Selected Central European Countries.* Food and Agriculture Organization of the United Nations, Rome.

Phillips, R. (1996) Communism strikes back: cultural blockages on the road to reform in the post-Soviet tourism sector. In: Robinson, M., Evans, N. and Callaghan, P. (eds) *Tourism and Culture: Image, Identity and Marketing.* Business Education Publishers, Sunderland, UK, pp. 147–164.

Phinnemore, D. (1999) The challenge of EU enlargement: EU and CEE perspectives. In: Henderson, K. (ed.) *Back to Europe: Central and Eastern Europe and the European Union.* UCL Press, London, pp. 71–88.

Pickel, A. (1997) Creative chaos: concluding thoughts on interdisciplinary cooperation. In: Jarausch, K. (ed.) *After Unity: Reconfiguring German Identities.* Berghahn, Oxford, pp. 196–207.

Pickel, A. (1998) Is Cuba different? Regime stability, social change, and the problem of reform strategy. *Communist and Postcommunist Studies* 31(1), 75–90.

Popesku, J. (2000) *Spas and Health Resorts in Serbia.* National Tourism Organization of Serbia, Belgrade.

Qiao, Y. (1995) Domestic tourism in China: politics and development. In: Lew, A.A. and Yu, L. (eds) *Tourism in China: Geographical, Political and Economic Perspectives.* Westview Press, Boulder, Colorado, pp. 121–130.

Quang, N. and Kammeier, H.D. (2002) Changes in the political economy of Vietnam and their impacts on the built environment of Hanoi. *Cities* 19(6), 373–388.

Riley, R.C. (2000) Embeddedness and the tourism industry in the Polish Southern Uplands: social processes as an explanatory framework. *European Urban and Regional Studies* 7(3), 195–210.

Roberts, L. and Hall, D. (2001) *Rural Tourism and Recreation: Principles to Practice.* CAB International, Wallingford, UK.

Robson, E. (2002) A nation of lace workers and glassblowers? Gendering of the Maltese souvenir handicraft industry. In: Swain, M.B. and Momsen, J.H. (eds) *Gender/Tourism/Fun(?).* Cognizant Communication Corporation, New York, pp. 109–117.

Rudolph, H., Appelbaum, E. and Maier, F. (1994) Beyond socialism: the uncertain prospects for East German women in a unified Germany. In: Aslanbeigui, N., Pressman, S. and Summerfield, G. (eds) *Women in the Age of Economic Transformation: Gender Impact of Reforms in Post-socialist and Developing Countries.* Routledge, London, pp. 11–26.

Rugg, D.S. (1971) Aspects of change in the landscape of East-central and Southeast Europe. In: Hoffman, G.W. (ed.) *Eastern Europe: Essays in Geographical Problems.* Methuen, London, pp. 83–126.

Rugg, D.S. (1985) *Eastern Europe.* Longman, London.

Rugg, D.S. (1994) Legacies of communism in the Albanian landscape. *Geographical Review* 84(1), 69–73.

Ryan, M. (2002) 'Axis of evil': the myth and reality of US intelligence and policy-making after 9/11. *Intelligence and National Security* 17(4), 55–76.

Saiko, T.A. (1998) Environmental challenges in the new democracies. In: Pinder, D. (ed.) *The New Europe: Economy, Society and Environment.* John Wiley & Sons, Chichester, UK, pp. 381–399.

Saltmarshe, D. (2000) Identity in a Post-communist Balkan State: a Study in North Albania. Unpublished PhD thesis, University of Bath, Bath, UK.

Schäfer, R. (2001) Frauen-Rechtsorganisationen in Südafrika: Ansätze und Grenzen gesellschaftlicher Transformationsprozesse. *Afrika Spectrum* 36(2), 203–222.

Scott, J. (1995) Sexual and national boundaries in tourism. *Annals of Tourism Research* 22(2), 385–403.

Scott, J. (1997) Chances and choices: women and tourism in northern Cyprus. In: Sinclair, M.T. (ed.) *Gender, Work and Tourism.* Routledge, London, pp. 60–90.

Seaton, A.V. (1996) Tourism and relative deprivation: the counter-revolutionary pressures of tourism in Cuba. In: Robinson, M., Evans, N. and Callaghan, P. (eds) *Tourism and Culture: Image, Identity and Marketing.* Business Education Publishers, Sunderland, UK, pp. 197–216.

Seaton, A.V. (1997) Demonstration effects or relative deprivation? The counter-revolutionary pressures of tourism in Cuba. *Progress in Tourism and Hospitality Research* 3, 307–320.

Segre, R., Coyula, M. and Scarpaci, J.L. (1997) *Havana: Two Faces of the Antillean Metropolis.* John Wiley & Sons, Chichester, UK.

Sharpley, R. (2001) The challenge of developing rural tourism in established coastal destinations: lessons from Cyprus. In: Mitchell, M. and Kirkpatrick, I. (eds) *New Directions in Managing Rural Tourism: Local Impacts, Global Trends.* Scottish Agricultural College, Auchincruive, CD-ROM.

Sharpley, R. (2003) Tourism in Cyprus: challenges and opportunities. *Tourism Geographies* 3(1), 64–86.

Siiskonen, P. (1996) *Overview of the Socio-economic Position of Rural Women in Selected Central and Eastern European Countries.* Food and Agriculture Organization of the United Nations, Rome.

Simon, F.L. (1995) Tourism development in transition economies – the Cuba case. *Columbia Journal of World Business* 30(1), 26–40.

Slovenia Tourist Board (1998) *Slovenia Tourist Board: the Role of STB.* Slovenia Tourist Board, Ljubljana <http://www.tourist-board.si/vloga-eng.html>.

Slovenia Tourist Board (1999) *Slovenia at a Glance: Some Brief Notes for Press Visitors.* Slovenia Tourist Board, Ljubljana <http://www.slovenia-tourism.si/enews/article-01.html>.

Smith, M. (2000) Negotiating new Europes: the roles of the European Union. *Journal of European Public Policy* 17(5), 806–822.

Sofield, T.H.B. and Li, F.M.S. (1998) Tourism development and cultural policies in China. *Annals of Tourism Research* 25(2), 362–392.

Staddon, C. (1999) Localities, natural resources and transition in Eastern Europe. *Geographical Journal* 165(2), 200–208.

Staddon, C. (2001) Local forest-dependence in postcommunist Bulgaria; a case study. *GeoJournal* 55(2-4), 517–528.

Stark, D. (1992) Path dependence and privatization strategies in East Central Europe. *European Politics and Societies* 6(1), 17–54.

Stark, D. and Bruszt, L. (1998) *Postsocialist Pathways*. Cambridge University Press, Cambridge.

Stifanic, D. (2000) *Vrsar: Bike Eco Ride*. Tourist Association of Vrsar, Vrsar.

Swain, M.B. (1995) Gender in tourism. *Annals of Tourism Research* 22(2), 247–266.

Thomas, M. (2002) Out of control: emergent cultural landscapes and political change in urban Vietnam. *Urban Studies* 39(9), 1611–1624.

Tomer, J.F. (2002) Intangible factors in East European transition: a socio-economic analysis. *Post-Communist Economies* 14(4), 421–444.

Transun (1998) *Transun's Croatia*. Transun, Oxford.

Uhlenbruck, K., Meyer, K.E. and Hitt, M.A. (2003) Organizational transformation in transition economies: resource-based and organizational learning perspectives. *Journal of Management Studies* 40(2), 257–282.

United Nations (2002) *International Migration from Countries with Economies in Transition: 1980–1999*. Department of Economic and Social Affairs, Population Division, United Nations Secretariat, New York <http://www.un.org/esa/population/publications/ewmigration/E-W_Migrationreport.pdf >.

Verbole, A. and Mele-Petric, M. (1996) *National Action Plan for Integration of Farm and Rural Women in Development: a Case Study of Slovenia*. FAO Workshop, Wageningen.

Warner, J. (1999) North Cyprus: tourism and the challenge of non-recognition. *Journal of Sustainable Tourism* 7(2), 128–145.

Weiner, E. (1997) Assessing the implications of political and economic reform in the post-socialist era: the case of Czech and Slovak women. *East European Quarterly* 31(3), 473–502.

Wen, Z. (1997) China's domestic tourism: impetus, development and trends. *Tourism Management* 18(8), 565–571.

Wieczorek, E. (1999) *The Silesian Voivodship Invites: Active Tourism*. Silesian Voivodship Tourist and Promotion Office, Katowice.

Williams, A.M. and Baláž, V. (2000) *Tourism in Transition: Economic Change in Central Europe*. Tauris, London.

Williams, A.M. and Baláž, V. (2001) Domestic tourism in a transition society: from collective provision to commodification. *Annals of Tourism Research* 28(1), 27–49.

Williams, A.M. and Baláž, V. (2004) From state socialism to EU membership: reflections on the transformation of tourism in Slovakia. *International Journal of Tourism Research* 6(2) (in press).

Williams, R.H. and Bjarnadottir, H. (1998) Environmental protection and pollution control in the EU. In: Pinder, D. (ed.) *The New Europe: Economy, Society and Environment*. John Wiley & Sons, Chichester, UK, pp. 401–413.

Witak, A. and Lewandowska, U. (eds) (1996) *Poland: the Natural Choice*. Sport i Turystyka, Warsaw.

WTO (World Tourism Organization) (1997) *The World's Most Important Tourist Destinations in 2020*. WTO, Madrid <http://www.world-tourism.org/calen.htm>.

Wu, Y.S. and Sun, T.W. (1998) Four faces of Vietnamese communism: small countries' institutional choice under hegemony. *Communist and Post-Communist Studies* 31(4), 381–399.

Yeonchul, K. (2001) Pragmatism or intransigence? Cuban lessons for North Korean economic reform. *East Asian Review* 13(3), 105–119.

Young, C. and Kaczmarek, S. (1999) Changing the perception of the post-socialist city: place promotion and imagery in Łódź, Poland. *Geographical Journal* 165(2), 183–191.

Zetter, R. (1994) The Greek-Cypriot refugees: perceptions of return under conditions of protracted exile. *International Migration Review* 28(2), 307–322.

Zhang, H.Q. and Qu, H. (1996) The trends of China's outbound travel to Hong Kong and their implications. *Journal of Vacation Marketing* 2(4), 373–381.

3 Capital accumulation – tourism and development processes in Central and Eastern Europe

Lesley Roberts

Introduction

Since the fall of state socialism, the transforming economies of Central and Eastern Europe (CEE) have provided rich yet relatively minor subjects for development studies. Historically, relatively little attention has been paid to the region as a whole in terms of its development (Nove *et al.*, 1982: 253), and the comparative neglect of its current tourism development would appear to mirror this indifference (Hall, 1991: 3, 1998a: 347). Moreover, the empirical nature of much extant research reflects a lack of theoretical framework with which to begin to understand complex development processes (Williams and Baláž, 2002). Yet the region has been in receipt of billions of dollars of overseas assistance, comprising grant aid, loans and other resources. Between 1990 and 1996 alone, G-24 commitments to CEE totalled US$80.823 million equivalent (Wedel, 1998: Appendix 1). To this must be added contributions from multinational agencies such as the World Bank, IMF, USAID, EU PHARE, EBRD, EIB and the UK Know-How Fund. Total net receipts of external resources to the region between 1990 and 1995 exceeded US$134 million (Financial Times, 1994; OECD, 1998b). The distribution of such funds, however, has been uneven, and reviews and discussions regarding criteria for disbursement and their efficacy rarely appear in tourism literature (Hall,

1995: 369), and are not generally available in the public domain (Lindberg *et al.*, 2001).

Indeed, as arguably the world's largest industry, tourism has a relatively low profile in international development processes generally, and there has been little work connecting the two fields of study (Sharpley and Telfer, 2002: 1). Although tourism has benefited indirectly from non-tourism funded programmes, it is both under-valued and under-exploited as a development tool in its own right (Steck *et al.*, 1999). Within the international development literature, where tourism is acknowledged as a contributory factor, its focus is often limited to the achievement of goals dealing with the conservation of natural and cultural heritage (Lindberg *et al.*, 2001). Within the tourism literature, tourism's development capacity is largely treated as an economic issue. For example, Butler's (1980) lifecycle model focuses on the industrial development of mass tourism and related economies of scale for resource management purposes. Otherwise, tourism's remaining potential has become absorbed into a broad discourse on sustainability that locates it in a quite different debate. Within the broader field of development studies, the agenda has evolved from narrow economic perspectives to embrace those of welfare, equity, choice and participation, as well as access to knowledge and the capacity to enjoy it (UNDP, 1990; Sharpley, 2002). Despite recent additions to the tourism literature (Milne and Ateljevic, 2001; Sharpley and Telfer, 2002), tourism has not yet received the attention it deserves by virtue of its contribution to the development of global capital (see Chapter 1).

The question of what constitutes 'capital' thus becomes a key question, mirroring longstanding debate in economic theory that now spills over into the development field and thereby into issues of tourism and development. Capital can be identified in a number of forms: financial, physical, natural, human and social, and it is generally understood that strength across the range is required for sustained economic growth and development. Nevertheless, critical differences exist between the dimensions, and each contributes to development with varying capacities.

Financial and physical capital are often private goods, tangible assets that can be generated, stored, used, devalued and destroyed. Human capital, both a private and a public good, can also be consciously created and is subject to a range of influences, both tangible and intangible, that may harness its use to varying degrees. Stocks of such kinds of capital help to explain economic growth paths of regions and of countries (Paldam and Svendsen, 2002), and it is relatively easy to conceptualize and measure their contributions to development processes. Conversely, social capital is more problematical, for whilst its contribution to development is increasingly recognized, its multiplicity of meanings lessens its explanatory powers (Woolcock, 1998) and renders measurement of it complex. Variously defined as 'density of trust' (Paldam and Svendsen, 2000), 'social capability' (Koo and Perkins, 1995) and 'the ability of people to work together for common purposes in groups and organizations' (Coleman, 1988: 95), social capital has no universally accepted definition. Rather, analysts tell us what the concept involves. Fundamentally it relates to social structures, collaboration, understandings of reciprocity and trust, and networks and institutions that contribute to development and growth. Social capital is believed to be a predictor of both individual income and macroeconomic growth in most countries (Bjørnskov, 2002). At the level of the individual, social capital reflects a person's ability to generate and maintain social connections and networks; in this respect it is a private good. At the level of

society, the focus of most development interest (and indeed of this chapter), social capital as a public good is about collective action, social participation, and the number, nature and quality of groups that constitute governance (Matějů, 2002). However, although the vital consequences of social capital are recognized, its sources are not, and its means of production have become the Holy Grail of the development fraternity.

Development and CEE

Although development rates between former socialist states of CEE vary somewhat, it is possible to identify two post-communist phenomena across the region. First, output fell more dramatically than was forecast; second, recovery was slower. Expectations had been calculated against valuations of tangible assets such as human and physical capital (Paldam and Svendsen, 2002), and given the low standards of living, it would not have been unreasonable to expect that injections of financial capital would create swift economic growth. Yet some essential ingredient(s) eluded transition processes, and a plethora of theories appeared by way of explanation. Catalysts to development in the form of external consultants cited lack of business knowledge, technology, and entrepreneurialism (Simpson and Roberts, 2000), while others suggested that social capital was the 'missing link' (Paldam and Svendsen, 2002). Without explicit reference to the concept of social capital, other studies, of partnership working for example, have identified factors critical to development success as: inclusion, trust, reciprocity, interdependence and legitimacy (Jamal and Getz, 1995; Roberts and Simpson, 1999a). The importance of networking and collaboration is stressed in tourism development studies. Thus it is possible to see that development studies from a number of perspectives support the idea that concepts related to social capital are essential ingredients of development economics.

Tourism's Role in Development Processes

Throughout CEE, since 1989, tourist agencies, new regional government organizations, non-governmental organizations, and community groups have emerged in response to potential growth in new tourism industries and the requirements of donor-funded programmes, thus expanding the numbers of stakeholders involved in tourism. The stakeholder society, a recent corrective to free-market capitalism in the West (Robson and Robson, 1996), provides a platform for the articulation of needs by a number of groups in the interests of consensus (Roberts and Simpson, 1999a). This echoes the principles of participative planning that aim to decentralize decision making, integrating it into community objectives, a process that requires such wider stakeholder involvement (Simmons, 1994).

Where social capital is well developed, groups share norms and values, cooperate voluntarily, and are able to subordinate individual group interests to those of the wider community. Jamal and Getz (1995) identified six factors critical to successful partnership working, including: stakeholder inclusion, recognition of interdependence and mutual benefits, the establishment of clear aims and objectives and of a legitimate convenor, and trust that decisions arrived at will be implemented, but the means by which such conditions may be achieved are uncertain. In their comparison of emerging tourism institutions in Bulgaria and Romania, Roberts and Simpson (1999b) identified that issues such as lack of trust, and failure to appreciate the reciprocal and common benefits derived from collaborative working, were barriers to effective development.

Certainly, development generally, and tourism development in particular, across the CEE region vary both within and between countries despite the considerable inputs of financial and physical capital and resources. There may be a number of reasons for this. For example, Williams and Baláž (2002) drawing on the work of Stark (1992, 1994) and others (Grabher and Stark, 1997; Stark and Bruszt, 1998), cite *path dependency* as one (see Chapter 2), referring to development constraints imposed by existing (and often entrenched) institutional resources that have created continuity rather than allowing change in tourism in the region. Paldam and Svendsen (2002) explain how the actions of more totalitarian regimes resulted in an atomization of social relationships, on the one hand suppressing positive forms of social capital and minimizing chances of its future development, whilst on the other generating negative forms of social capital through the promulgation of fear and distrust. This suggests that the determinants of social capital vary, with implications for policy formulation.

Development Assistance in the CEE Region

Since the collapse of the former communist regimes, many international aid organizations have offered assistance to the restructuring economies of the region to ease transitions and support moves towards more stable economies and democracies. USAID, the World Bank, the EU, and the UK Know-How Fund, amongst others, have funded and delivered development programmes representing significant flows of funds. Most programmes share common aims including capacity building for civil society, the promotion of democratic principles, and economic restructuring and development. In their turn, CEE country governments have had the considerable task of developing new administrations, policies and strategies, and processes to support such change. With the loss of traditional markets and the subsequent decline of associated manufacturing sectors, tourism development appears high on the list of priority sectors for development for most CEE governments where it is seen as a vehicle for developing local, regional and national economies (Hall, 1991, 1995, 1998b; Fletcher and Cooper, 1996). Within development processes, although financial capital has been central to project planning, consultants have played a pivotal role in helping to shape the nature of developments, particularly in relation to institutional capacity building and human resource development in order to encourage grassroots development, partnership working, and local ownership and control through participation (Simpson and Roberts, 2000). Given the complexities of defining and understanding social capital, and the scale and importance of such a remit for development practitioners, it is not difficult to see why so many projects have underestimated the capital inputs required and have failed to achieve their aims. The considerable criticism that has ensued (Streeten, 1995; Cowen and Shenton, 1996; Preston, 1997; Randel and German, 1998; Wedel, 1998), serves only to reinforce the importance of intangible elements of development, and to create a demand for more focus on the ways in which project aims might be achieved through social capital building. In particular, the processes by which change is sought and engendered, where they attempt to influence social capital, are of interest; the challenge to policy makers and development workers is to identify deficits in social capital in the interests of new social relationships that foster positive forms of community development.

The Economic Significance of Tourism in CEE

Increases in real incomes and in leisure time within the world's tourism-generating countries have led to increased demands for leisure, recreation and tourism. Predictions by the World Tourism Organization (WTO) suggested increases in tourist arrivals globally of 4.1% per annum between 1996 and 2006 (WTO, 1999). Growth for Europe as a whole is predicted to be less, at 3.5%, as the region loses out to newly emerging destinations, particularly in Asia and the Pacific (Pearce, 1999: 1). For the Central and Eastern European region, however, the WTO predicts growth in arrivals ahead of both the European and world averages at an annual average of 4.6% (TTI, 2000: 220). Tourism is an increasingly important element of the CEE economy. As a proportion of GNP (a potential indicator of the growing economic importance of tourism since 1989) tourism receipts overtook the European average as far back as 1994 (Hall, 1998a: 359).

The economic value of tourism may be measured by contribution to balance of payments, income, and/or GDP and by job creation potential. For reasons of simplicity and measurability, estimates of direct tourist expenditure are often used to construct the value of tourism to a country's economy (Williams and Shaw, 1994: 4).

International tourist arrivals and tourism receipts in CEE for the 10-year period 1989–1998 are shown in Table 3.1 below (see also Tables 1.2, 1.3 and 15.1).

Receipts from tourism showed dramatic increases in the mid-1990s, although not to the extent predicted (TTI, 2000: 1), but have now slowed considerably. The pace of economic reforms, continued political instability, and security issues on the supply side, together with slowing economic growth and greater choice in source markets, may help to explain why predicted levels of growth were not achieved.

Evaluations and assessments of volumes, values and patterns of tourism are complicated by problems associated with the collection and presentation of data. As pointed out by a number of writers (Hall, 1991, 1998a; Witt, 1994; Vospitannik et al., 1997), there are inconsistencies surrounding data quality and availability for the region as a whole as well as for constituent countries. Caution therefore needs to be exercised in their interpretation. From that point of view alone it is very difficult to analyse issues of financial capital as they relate to the economic importance of any industrial sector. Although clearer pictures emerge as institutions stabilize and data collection systems standardize and mature, conclusions cannot be drawn without reservation. Nevertheless, despite the fact that tourism is subject to two major (and linked) processes in addition to international pressures – privatization of property rights, and inward investment (Williams and Shaw, 1998: 7) – it is possible to identify long-term regional growth in international tourism in CEE (Hall, 1998a: 358; TTI, 2000). The enhancement of such potential through the accumulation of capital inputs further highlights the need to focus on issues of social capital, particularly in rural areas on the margins of restructuring processes.

Social Capital – the Missing Link?

Fundamental to much of what has been said so far is the belief that the slow pace of transition within many CEE countries is the result of low levels of (positive) social capital and its existence in negative form. If this is the case, it should be possible to make a direct compari-

Table 3.1. Central and Eastern Europe: international tourist arrivals and tourism receipts, 1989–1998.

	Arrivals ('000)	% annual change	Receipts (US$ million)	% annual change
1989	40,051	24.17	4,948	15.18
1990	43,809	9.38	4,849	−2.00
1991	52,703	20.30	7,474	54.13
1992	58,136	10.31	8,154	9.10
1993	66,081	13.66	9,972	22.30
1994	73,202	10.78	14,045	40.84
1995	75,565	3.23	18,191	29.52
1996	81,826	8.29	25,126	38.12
1997	79,480	−2.87	25,574	1.78
1998	76,535	−3.71	25,444	−0.51

Source: WTO, 2000.

son between East and West Europe in order to prove causality. As with measures of many intangible concepts such as perceptions or values, measures of social capital use proxies. Commonly, proxies are: civic engagement, interpersonal trust, confidence in government, and poverty (Hjøllund and Svendsen, 2004). Low social capital, for example, is measured as a low density of voluntary organizations, and it would appear that, for each of the above proxy measures, Eastern European countries fall behind their Western European counterparts. Nevertheless, simplicity constantly eludes the concept, and Bjørnskov (2002) has found that in Slovenia, for example, despite a density of voluntary organizations equating to Western European levels, levels of mutual trust, deemed to be central to social capital, remain low. In her case study of Pišeće, Slovenia, Alenka Verbole (2001) found that tourism development was dominated by strong historical social and kinship groups (illustrating *path dependency* in practice and the existence of negative social capital) and that inclusion in or exclusion from social organizational forms resulted directly from such group membership. Some organizational forms appeared to be quite arbitrary (if entrenched) with clear implications for sustainable tourism development. That which took place was shaped by local institutional practices with little evidence of trust, partnership or reciprocity. However, Verbole notes growth in the domain of voluntary organizations where membership is unrestricted, suggestive of a change in the nature of

social capital in the region and the potential for more inclusive development as a result.

In similar vein, Hall (2000) suggests that social relations in Albania resulting from atavistic customs concerned with family and clan honour have re-emerged as major constraints to development. Distrust of both neighbours and of government has created a vacuum within which negative forms of social capital have been able to thrive. Indeed, as pointed out by Bjørnskov (2002), trust in general is lower in southern Europe than in northern Europe, creating a double burden for postsocialist countries in the south of the region.

Exemplification from Romania

Assessments of Romania's progress in transitional processes paint a picture of procrastination, instability and inefficiency (TTI, 2000; OECD, 2002), suggestive of low social capital reserves. The new government is committed to acceleration of the reform process with two key drivers – EU and NATO accession – but a great deal has yet to be achieved in a short time, and on the back of long delays. The OECD economic survey for 2002 makes many of the same points as in 1998 (OECD, 1998a), and a great deal of credibility has been lost by a gradualist approach. Moreover, international perceptions of corruption in the region are the second highest after Russia (Transparency International, 2001), suggesting loss of confidence by the international business

community. Nevertheless, despite the failure of the 1997 stabilization plan sufficiently to accelerate the privatization programme and promote foreign investment, the country's significant potential is still expected to be achieved.

In Romania, although almost €1 million were invested in rural tourism development through EU PHARE funding, the pilot programme, centred in Bran, did not achieve its potential (see Roberts, 1996; Roberts and Simpson, 1999a). Rivalry between national and international non-governmental organizations (NGOs) prevented the development of most of Jamal and Getz's (1995) requirements for successful partnership working. The evident lack of trust and of institutional partnership in this case would appear to support the theory that social capital and not financial capital is the factor critical to successful development. The availability of financial capital was not an obstacle, but the existence of social capital may

Fig. 3.1. Potential social capital? (source: Derek Hall). Opposite: (a) Croatia: cooperative fishermen, Rovinj. (b) Turkey: future tourism entrepreneurs? Orderly school children, Antalya. Above: (c) Albania: disorderly youths, Kukës, near the border with Kosovo. (d) Romania: cooperative farm women, Carpathian foothills. (e) Vietnam: likely young entrepreneurs, Hanoi.

continued

Fig. 3.1. *Continued.* (f) South Africa: roadside fruit sellers, Soweto.

have been. However, the situation in Bran must not be taken as a microcosm of Romanian development processes. Elsewhere in Romania, rural tourism is thriving, based on voluntarism. In Maramureş, a mountain region in the north, close to the Ukraine border, a further pilot project for rural tourism has been established. Unlike Bran, however, rural tourism in Maramureş has the realizable potential to become part of a programme of integrated regional reconstruction. International development aid from USAID is to support large tourism developments in Maramureş County while small-scale developments across a number of villages already exist within a network established with assistance from an international NGO. There is local municipal support for rural tourism development, and, at the village level, a number of NGOs and local associations have emerged that, together with cohesive village communities, reflect a growing range of stakeholders with interests in tourism development as part of integrated regional development. The Maramureş case (*see* Turnock, 2002) further illustrates the importance of social capital in development processes by presenting a multi-stakeholder society where partnership working and collaboration would appear to be providing added value to integrated development processes.

In their comparison of institutional support for the tourist industries of Romania and Bulgaria, Roberts and Simpson (1999b) examined three interlinked but distinct core issues affecting tourism development processes: the influence of donor support, institutional linkages in the locality, and community ownership and accountability. In Pirin, in south-west Bulgaria, an environmental NGO's status within a donor-supported tourism development project was seen as an advantage in developing perceptions of impartiality and in the generation of trust in the organization and its work. The NGO was thus able to build an important synergistic role by bringing together otherwise marginal institutions. Furthermore, its neutral role was pivotal in coordinating the work of local municipalities, a critical issue in post-socialist states where effective NGO-government partnerships can help to re-establish severed links between the state and the individual.

Across CEE as a whole, constraints on tourism development have tended to reflect obstacles facing development processes in general, such as procrastination, instability and inefficiency. In his study of Albania, Hall (2000) points out that the lack of stability in communities in transition may undermine participation in development processes and that models of community development are incapable of accommodating instability. This suggests a need for more flexible development processes able to reflect difficult and changing contexts.

Conclusions

As socialist, centrally planned regimes fell, and the East and West looked each to the other for the means by which 'democracy' and market economics might be achieved in the area, financial capital was the major currency for exchange. With the experience of almost 15 years of transition, however, it would appear that financial capital is neither the only nor even the main constraint on development processes, and that social capital may provide a more useful focus of study. However, this is problematic for a number of reasons. First, if we cannot agree on common definitions and understandings of the term, it remains nothing more than an elusive and intangible concept incapable of measurement. Second, as such, it cannot be operationalized in development processes. Third, the existence of path dependency lengthens the time taken to develop the trust that underpins willing cooperation and reciprocity. Finally, by definition, the development of social capital is a voluntary activity, and any attempts, either by the state or by 'development workers' formally to induce cooperation at once remove its very essence – its voluntarism.

The conclusion that social capital carries such importance in the development of tourism presents developers with the conundrum of how it is to be created. The contributions of international aid agencies carry donor requirements and the involvement of aid workers and/or consultants external to cultural contexts. The promotion of voluntarism is an issue for reflective practice (Schon, 1983), requiring a reactive rather than a proactive approach to external consultancy practice (Simpson and Roberts, 2000). As observed by Roberts and Simpson (1999a), it is the less tangible factors – which are the more critical to the success of collaborative working – that appear to be both fundamental to and a measure of the existence of social capital. Recognition of interdependence and of mutual benefits may help to build trust; the existence of transparency will further support its development.

Financial capital will continue to be required and this too may not be without its problems. Although EU PHARE and TEMPUS programmes, for example, have supported schemes for education and training in tourism across CEE, there is limited evidence to indicate the long-term viability of tourism outside key urban and resort areas. Rather, investment and entrepreneurial activity have been concentrated in larger urban zones (Hall, 1998a: 353).

Capital flows of all kinds are required for more sustainable tourism development, and issues of capital accumulation, being addressed by development theory, must be added to the tourism development agenda.

References

Butler, R.W. (1980) The concept of a tourist area lifecycle of evolution and implications for management of resources. *The Canadian Geographer* 24 (1), 5–12.

Bjørnskov, C. (2002) Policy Implications of Social Capital in Post-Communist Europe. Is Slovenia Different from the Rest? Paper presented at *Institutions in Transition* conference, Ljubljana, Slovenia, 23–24 September.

Coleman, J. (1988) Social capital in the creation of human capital. *American Journal of Sociology* 94, 95–120.

Cowen, M.P. and Shenton, R.W. (1996) *Doctrines of Development*. Routledge, London.

Financial Times (1994) *East European Markets*. Financial Times, London.

Fletcher, J. and Cooper, C. (1996) Tourism strategy planning, Szolnok County, Hungary. *Annals of Tourism Research* 23(1), 181–200.

Grabher, G. and Stark, D. (eds) (1997) *Restructuring Networks in Postsocialism: Legacies, Linkages and Localities.* Oxford University Press, Oxford.

Hall, D. (1991) Introduction. In: Hall, D. (ed.) *Tourism and Economic Development in Eastern Europe and The Soviet Union.* Belhaven Press, London, pp. 3–28.

Hall, D. (1995) Tourism change in Central and Eastern Europe. In: Montanari, A. and Williams, A.M. (eds) *European Tourism, Regions, Spaces and Restructuring.* John Wiley & Sons, Chichester, UK, pp. 221–244.

Hall, D. (1998a) Central and Eastern Europe: tourism, development and transformation. In: Williams, A.M. and Shaw, G. (eds) *Tourism and Economic Development: European Experiences.* John Wiley & Sons, Chichester, UK, pp. 345–373.

Hall, D. (1998b) Tourism development and sustainabiilty issues in Central and South-eastern Europe. *Tourism Management* 19(5), 423–431.

Hall, D. (2000) Tourism as sustainable development? The Albanian experience of 'transition'. *International Journal of Tourism Research* 2, 31–46.

Hjøllund, L. and Svendsen, G. (2004) Social capital: a standard method of measurement. In: Paldam, M. and Svendsen, G. (eds) *Trust, Social Capital and Economic Growth: An International Comparison.* Edward Elgar, Cheltenham, UK.

Jamal, T. and Getz, D. (1995) Collaboration theory and community tourism planning. *Annals of Tourism Research* 22(1), 186–204.

Koo, B.H. and Perkins, D. (1995) *Social Capability and Long-term Economic Growth.* St Martin's Press, New York.

Lindberg, K., Andersson, T. and Dellaert, B. (2001) Tourism development, assessing social gains and losses. *Annals of Tourism Research* 28(4), 1010–1030.

Matějů, P. (2002) Social Capital: Problems of its Conceptualisation and Measurement in Transforming Societies. Paper presented at the OECD-ONS Conference on *Social Capital Measurement,* London, 26–27 September.

Milne, S. and Ateljevic, I. (2001) Tourism, economic development and the global-local nexus: theory embracing complexity. *Tourism Geographies* 3(4), 369–393.

Nove, A., Hohman, H. and Seidenstecher, G. (1982) *The East European Economies in the 1970s.* Butterworth, London.

OECD (1998a) *OECD Economic Surveys: Romania, Economic Assessment.* OECD, Paris.

OECD (1998b) *Aid and Other Resource Flows to the Central and Eastern European Countries and the Newly Independent States of the Former Soviet Union.* Working Paper IV (3), OECD, Paris.

OECD (2002) *OECD Economic Surveys: Romania, Economic Assessment.* OECD, Paris.

Paldam, M. and Svendsen, G.T. (2000) An essay on social capital: looking for the fire behind the smoke. *European Journal of Political Economy* 16, 339–366.

Paldam, M. and Svendsen, G.T. (2002) Missing social capital and the transition in Eastern Europe. *Journal for Institutional Innovation, Development and Transition* 5, 21–34.

Pearce, D. (1999) Introduction. In: Pearce, D. and Butler, W. (eds) *Contemporary Issues in Tourism Development.* Routledge, London, pp. 1–12.

Preston, P.W. (1997) *Development Theory: An Introduction.* Blackwell, Oxford.

Randel, J. and German, T. (eds) (1998) *The Reality of Aid: An Independent Review of Poverty Reduction and Development Assistance.* Earthscan, London.

Roberts, L. (1996) Barriers to the development of rural tourism in the Bran area of Transylvania. In: Robinson, M., Evans, N. and Callaghan, P. (eds) *Tourism and Culture: Image, Identity and Marketing.* Business Education Publishers, Sunderland, UK, pp. 185–196.

Roberts, L. and Simpson, F. (1999a) Developing partnership approaches to tourism in Central and Eastern Europe. *Journal of Sustainable Tourism* 7(3/4), 314–330.

Roberts, L. and Simpson, F. (1999b) Institutional support for the tourist industry in post-socialist Europe: a comparison of Bulgaria and Romania. *Tourism Recreation Research* 24(2), 51–58.

Robson, J. and Robson, I. (1996) From shareholders to stakeholders: critical issues for tourism marketers. *Tourism Management* 17(7), 533–540.

Schon, D. (1983) *The Reflective Practitioner. How Professionals Think in Action.* Basic Books, New York.

Sharpley, R. (2002) Tourism, a vehicle for development? In: Sharpley, R. and Telfer, D. (eds) *Tourism and Development, Concepts and Issues.* Channel View Publications, Clevedon, UK, pp. 11–34.

Sharpley, R. and Telfer, D. (2002) (eds) *Tourism and Development, Concepts and Issues.* Channel View Publications, Clevedon, UK.

Simmons, D.G. (1994) Community participation in tourism planning. *Tourism Management* 15(2), 98–108.

Simpson, F. and Roberts, L. (2000) Help or hindrance? Sustainable approaches to tourism consultancy in Central and Eastern Europe. *Journal of Sustainable Tourism* 8(6), 491–509.

Stark, D. (1992) Path dependence and privatization strategies in East and Central Europe. *East European Politics and Society* 6(1), 17–54.

Stark, D. (1994) Path dependence and privatization strategies in East and Central Europe. In: Miloy, E. (ed.) *Changing Political Economies – Privatization in Post-communist and Reforming Communist States.* Lynne Reiner, New York, pp. 116–148.

Stark, D. and Bruszt, L. (1998) *Postsocialist Pathways.* Cambridge University Press, Cambridge.

Steck, B., Strasdas, W. and Gustedt, E. (1999) *Tourism in Technical Co-operation: a Guide to the Conception, Planning and Implementation of a Project.* GTZ and BMZ, Germany.

Streeten, P. (1995) *Thinking About Development.* Cambridge University Press, Cambridge.

Transparency International (2001) *New Europe Barometer 2001.* Centre for the Study of Public Policy, London.

TTI (2000) *Tourism in Central and Eastern Europe.* Travel and Tourism Intelligence, London.

Turnock, D. (2002) Prospects for sustainable rural cultural tourism in Maramureş, Romania. *Tourism Geographies* 4(1), 62–94.

UNDP (1990) *Human Development Report.* United Nations, New York.

Verbole, A. (2001) Networking and partnership building for rural tourism: politics of rural tourism in Slovenia. In: Mitchell, M. and Kirkpatrick, I. (eds) *New Directions in Managing Rural Tourism and Leisure: Local Impacts, Global Trends.* Scottish Agricultural College, Auchincruive, UK, CD-ROM.

Vospitannik, N., Littlejohn, D. and Arnot, R. (1997) Environments, tourism and tour operators; 1985–1995 in Central and Eastern Europe. *International Journal of Contemporary Hospitality Management* 9(5/6), 209–214.

Wedel, J. (1998) *Collision and Collusion, the Strange Case of Western Aid to Eastern Europe 1989–1998.* Macmillan, Basingstoke, UK.

Williams, A.M. and Baláž, V. (2002) The Czech and Slovak Republics: conceptual issues in the economic analysis of tourism in transition. *Tourism Management* 23, 37–45.

Williams, A.M. and Shaw, G. (eds) (1994) *Tourism and Economic Development: Western European Experiences,* 2nd edn. Belhaven Press, London.

Williams, A.M. and Shaw, G. (eds) (1998) *Tourism and Economic Development: European Experiences.* John Wiley & Sons, Chichester, UK.

Witt, S. (1994) Opening of the Eastern Bloc countries. In: Theobald, W. (ed.) *Global Tourism, the Next Decade.* Butterworth Heinemann, Oxford, pp. 217–225.

Woolcock, M. (1998) Social capital and economic development: towards a theoretical synthesis and policy framework. *Theory and Society* 27, 151–208.

WTO (1999) *Compendium of Tourism Statistics.* World Tourism Organization, Madrid.

WTO (2000) *Yearbook of Tourism Statistics.* World Tourism Organization, Madrid.

4 The Government's role in stimulating national tourism development: the case of Poland

Wacław Kotlinski

Introduction

Systemic changes took place in Poland in the 1980s, as elsewhere in Central and Eastern Europe, which led to the development of a market economy with all its characteristic features: privatization of state assets, growth of private small businesses, an open economy with foreign investors and multinational penetration, orientation of enterprises towards gaining profits, and other market mechanisms (Gołembski, 1990; Dawson, 1991).

The economic and political crisis which stimulated political change in 1989 also affected the tourist trade. Socialist governments had treated tourism as an element of social politics at a domestic level, and as a field under special supervision for 'privileged people' in relation to foreign tourism. In the state socialist period, Polish tourism was characterized by the following features (Gralec, 1996):

- centralized management and financing;
- division into social tourism (subsidized for the domestic but mostly urban population) and market tourism;
- tourism management was subordinate to social rather than market criteria: as a consequence there was lack of correlation between the price of tourist services and the expenditures which were paid in order to create the service – the quality of services suffered as a result;

- state ownership of the tourism infrastructure was dominant; and
- constraints on the development of foreign tourism, both inbound and outbound, which manifested themselves in such administrative impositions as passport and visa difficulties, and restrictions on the possession and spending of foreign currencies by Polish nationals (Gołembski, 1990; Dawson, 1991).

Political and economic change encouraged a rapid development of tourism, and consequently of the income which flowed from tourism. Such income began to constitute a substantial contribution to the gross national product (GNP) of Poland. Further economic development of the country calls for a strategy targetting particular economic sectors to fulfil the role of growth stimulators. The aim of this chapter is to explore whether tourism can play the role of such a stimulator, and if so, what actions should be expected from the government to develop the tourist economy in Poland.

Characterization of the State of Tourism in Poland

Poland has been recognized by the World Travel Organization (WTO) as 'a tourist tiger' (Migdal, 1999). Between 1990 and 1998 the number of international tourists grew by 550% from 3.4 million to 18.8 million, and Poland rose in WTO rankings from 27th to 8th most-visited country. Receipts grew even more spectacularly by about 2000% from US$358 million to US$8.0 billion, whereby Poland moved from 64th to 12th ranking destination. These trends have reversed notably, however, since 1998 (Tables 4.1 and 4.2).

The dynamic development of the tourist trade, which has been observed in Poland since the beginning of the processes of transformation of the economy, is also rooted in the country's transformations, which have radically changed the Polish tourist landscape. Crucial factors brought about by political change included an opening of international borders, annulment of foreign currency and visa restrictions, and a liberalization of the principles of business activity. The system of central author-

ity, the philosophy of the macroeconomic management and the management of tourist policies were all changed.

Tourism has become dominated by private firms, and former state and cooperative enterprises have undergone privatization processes, thereby restricting or eliminating their former social functions in favour of profit maximization.

Following a period of rapid development of tourism numbers and receipts, the latter years of the 1990s saw a notable reversal in fortunes. In the first half of the 1990s, the growth of tourist arrivals (Table 4.1) was influenced by the so-called 'cascade effect', relating to different levels of prices and availability of goods and services between neighbouring countries. In relation to Poland it caused a dynamic development of the border areas and gave positive effects to the balance of payments. Gradually, as a result of adaptation in both pricing and distribution processes in neighbouring countries, the significance of this factor has lessened for Poland (Szubert-Zarzeczny, 1999). This has resulted in new challenges for the Polish tourism industry and particularly for the national tourist office, in relation to the need for a review and modification of marketing strategies in response to the changing 'frame' conditions (Poon, 1993) of economic and non-economic factors influencing future directions of tourism development in Poland.

The level of foreign currency receipts in Poland has significance for the level of supply of money and for the origin of inflation impulses within the country. The growth of the

Table 4.1. Poland: arrivals of tourists and day visitors, 1994–2001.

Year	International tourist arrivals ('000s)	Total international arrivals ('000s)
1994	18.8	74.2
1995	19.2	82.2
1996	19.4	87.4
1997	19.5	87.8
1998	18.8	88.5
1999	18.0	89.1
2000	17.4	84.5
2001	15.0	61.4
2002	14.0	50.7

Source: Institute of Tourism, 2003.

Table 4.2. Poland: foreign currency receipts from incoming visitors, 1994–2001.

	Receipts in US$ billion		
Year	Tourism (involving overnight stay)	Day visitors	Total receipts
1994	3.6	2.5	6.1
1995	3.3	3.3	6.6
1996	4.4	4.0	8.4
1997	4.3	4.4	8.7
1998	4.4	3.6	8.0
1999	3.4	2.7	6.1
2000	3.1	3.0	6.1
2001	2.9	1.9	4.8
2002	2.8	1.7	4.5

Source: Institute of Tourism, 2003.

Fig. 4.1. Poland: Medieval Kraków in winter.

Fig. 4.2. Poland: Warsaw's old 'New' town.

foreign currency reserves from international tourism receipts means growth in the monetary base, in the supply of money and an increase in inflation. In the country's macroeconomic policy, particularly in the 1990s, tourist receipts were treated as an inflationary factor. This resulted from the fact that in Polish statistics about 97% of foreign tourists are registered in the balance of payments as short-term deposits, whereas usual international practice is to register tourism receipts as an export of services. This is therefore a fundamental difference in the attitude to the invisible tourist export between Poland and other countries. In Poland there is a lack of calculations, even estimates, which would allow an evaluation of the degree to which tourism receipts and capital movement contribute to foreign currency

Fig. 4.3. Poland: Hanseatic Gdansk.

growth, to national financial reserves, and to investment in Polish stocks. There is also a lack of accurate studies that refer to the scale of removal of foreign currency from Poland.

Between 1994 and 1997 there was continuous growth of international tourism receipts in global receipts of Poland, including the expenditures of day visitors (Table 4.2). This resulted from an abolition of restrictions on cross-border traffic and advantageous price levels of various Polish goods. Since 1998 there has been a permanent gradual decline in foreign currency receipts, particularly from day visitors. The government continued to maintain a higher and higher rate of the national currency, the *złoty*, which saw price differentiations between Poland and West European countries diminishing. This has had an unfavourable impact on the arrivals of day visitors, the overwhelming majority of whom have been petty traders from former Soviet territories.

Income derived from tourist spending constitutes an important source of finance for the social development and economic growth of recipient countries. The quotient of receipts from foreign tourism to the national revenue is the measurement which is applied here. Its value is modelled in a different way in different countries of tourist reception. Countries which have the highest level of revenues generated from foreign tourism rarely register a high-level contribution to GNP of foreign currency receipts from tourism (for example in 2000 the USA – 0.83%, Italy – 2.56%, France – 2.37%).

Other countries may accrue relatively low income from foreign tourism, but its GNP contribution is high (for example Greece – 8.2%, Austria – 5.3%, Switzerland – 3.1%). This results from the level of economic development of those countries and the place of tourism in the total trade turnovers of the given country.

In Poland, on account of very high receipts obtained from tourism, and on the other hand, the small value of the gross national product, the contribution of the receipts to the GNP is high (Table 4.3). The contribution to GNP amounted to 5.2% in 1995. Unfortunately, since 1996, when it was at its highest (6.2%), it has been undergoing a permanent decline, reaching a level of 3.9% in the year 2000. In spite of such a great decline, the receipts from tourism constitute a substantial contribution to the GNP of Poland.

The budget still remains the basic instrument of public financing in tourism. The role of the budget is systematically getting smaller and smaller (Table 4.4). The year 1998 was a high point when the government assigned most money, i.e. about US$19 million, to the development of tourism, but this constituted only 0.2% of the receipts obtained from tourism, in contrast to the 5.1% tourism contribution to GNP. These figures show how big the disproportions are between tourism receipts in Poland and government expenditures to support its development. After 1998 there was a decline both in government expenditure and in tourism receipts.

Experts from the Central Office for Planning argue that budget support of US$60–70 million has been necessary to realize the objectives of state tourist policy (for example the development of agrotourism, investments in the accommodation base) and in order to support social functions of tourism (for example

Table 4.3. Poland: contribution of tourism receipts to GNP, 1995–2000.

Years	Total visitor receipts (US$ million)	Total GNP (US$ million)	Contribution of receipts to GNP (%)
1995	6,600	126,348	5.22
1996	8,400	134,550	6.24
1997	8,700	143,066	6.08
1998	8,000	157,274	5.09
1999	6,100	155,151	3.93
2000	6,100	157,585	3.87

Source: CSO, 2002: 544.

Table 4.4. Poland: the relationship between tourism receipts and government expenditure for tourism, 1994–2000.

Years	Receipts from tourism (US$ million)	Government expenditure for tourism (US$ million)	Expenditure as a percentage of receipts
1994	6100	6.52	0.107
1995	6600	9.80	0.148
1996	8400	15.20	0.181
1997	8700	17.35	0.199
1998	8000	19.00	0.237
1999	6100	14.62	0.240
2000	6100	12.25	0.201

Source: CSO, 2002: 626–627.

school field courses, recreation for children and youth) (Central Office for Planning, 2000).

Factors Constraining the Development of Tourism in Poland

The recent decline in tourist arrivals to Poland requires an analysis of the factors constraining growth. These include (Gołembski, 1996):

- insufficient political support for tourism development, especially for new construction and modernization of the existing infrastructure;
- insufficient encouragement for foreign capital investment in tourism, and the creation of equal investment opportunities for both home and foreign capital;
- poorly developed capacity to permit successful management of the tourist economy at local and regional levels; and
- the need to better develop provision of tourist information and marketing in tourism, at both national and regional levels.

The construction and modernization of the infrastructure is a key issue conditioning the development of tourism. This is highlighted because the tourism industry base inherited from the previous political economic system is not equipped for increased quantity or improved quality. This results from the high degree of cumulative under-investment. It brings about the necessity to invest vast sums of money in order to undertake necessary improvements in the quality of the services to reach an acceptable standard. Financial requirements for reconstructing and developing the tourist base exceed the capacities of many investors in the contemporary economic climate. This results in delays in privatization processes, the collapse of many owners of the old base, a reduction and even the freezing of investment activity with Polish capital and an increase in hotel prices.

In the meantime there are almost no stimuli which would enable small service providers to undertake investment and repair activity. There is a noticeable absence of professional consulting institutes which would otherwise assist with an adjustment to the standard of services international tourists demand.

The development of the tourist base is connected with the influx of foreign capital. The engagement of this capital can be reflected both in direct investments (the construction of new tourist facilities) and in indirect investments (for example hotel acquisiton by foreign-owned chains). Unfortunately, this under-investment results from the observation that foreign capital is insufficiently interested in investing in Poland. The reasons for that lie both in the administrative legal sphere (the system of granting licences, methods of valuing property and land) and especially in the financial sector (the tax system, credit system, the opportunities for capital flow).

There exists an urgent need to change the flawed system of tourist management. The modelling of supply for tourist services must be supplemented substantially by funds coming from the national budget. Supply is also intimately connected with the tourist region and the coordinated management of numerous different components of this supply can be carried out only at the regional and local levels. With the present centralized system of budget accumulation and redistribution, an integrated and coherent supply of tourist services in the regions is not possible. Financial shortages in local budgets, a lack of correspondence between exchequer receipts from the tourism industry in a given region and returns from exchequer sources assigned for the development of tourism, are key elements which constrain the development of tourism in Poland. In addition to national sources of financing there are various types of assistance programmes from the European Union. However, in practice, access to those means is limited for local communities and small firms, both because of strict formal requirements for access and a lack of means for co-financing the investments. The consequence of this is that large amounts of European money, which were allocated to Polish investors, are not taken advantage of.

Conclusions

Despite a reduction in the number of tourist arrivals and level of receipts, tourism occupies a significant place in Poland's national economy. Acceptance of the assumption that tourism is a significant factor for national socio-economic development highlights the current insufficient exploitation of this factor in the stimulation of Polish economic growth as it enters a new stage of transformation. There is a strong requirement for substantial government action to establish the framework and create the appropriate conditions for the development of tourism in Poland.

For the further development of infrastructure, and particularly the accommodation base, action is required first of all in the tax sphere, in credit availability and in the formation of amortization. A matter of priority is the need for temporary exemption of tourist accommodation establishments from taxes in the period following investment and modernization. In the regions with greatest tourism potential, which at the same time are suffering from the scourge of unemployment, activities to support the development of private enterprises must be more dynamic and conducted on the basis of long-term benefits. Such support should embrace the assurance of long-term loans, tax exemptions, the establishment of funds for infrastructure development and the management of marketing on a regional scale.

Tourism in Poland loses its potential and synergies largely because of the dispersed and uncoordinated nature of attractions, and through the lack of an integrated system of management in the regions. The basic reason for this state of affairs is centralization of budget means (disproportion between receipts of the central budget and local budget) as well as lack of local self-government with a realistic possibility to take and to execute decisions. An absolute necessity is to establish a degree of hypothecation – to assign much more tax revenue derived from business activity in tourism to supply local budgets than has been the case hitherto.

Marketing and promotion of Poland abroad is the second priority for government activity. It should be conducted by the establishment of government institutions such as an office for the promotion of tourism, or other institutions of ministry status. The creation of a positive image of Poland as a country with a strong tourist appeal is a basic duty of those institutions.

References

Central Office for Planning (2000) *The Report of the Central Office for Planning 2000.* Central Office for Planning, Warsaw.

CSO (Central Statistical Office) (2002) *Statistical Yearbook of the Republic of Poland.* CSO, Warsaw.

Dawson, A.H. (1991) Poland. In: Hall, D. (ed.) *Tourism and Economic Development in Eastern Europe and the Soviet Union.* Belhaven, London, pp. 190–202.

Gołembski, G. (1990) Tourism in the economy of shortage. *Annals of Tourism Research* 17, 55–68.

Gołembski, G. (1996) Factors which block the development of tourism in Poland. In: *Current Issues of the Tourism Economy.* Academy of Economics, Poznań, pp. 87–99.

Gralec, R. (1996) Tourism in the conditions of the market economy. In: *Current Issues of the Tourism Economy.* Academy of Economics, Poznań, pp. 42–48.

Institute of Tourism (2003) *Tourism Sector – Data Analyses, Forecasts.* Institute of Tourism, Warsaw <http://www.intur.com.pl>.

Migdal, M. (1999) Perspectives of the development of tourism on the Western Pomerania. *Firm and Market* 13, 54–60.

Poon, A. (1993) *Tourism, Technology and Competitive Strategies.* CAB International, Wallingford, UK.

Szubert-Zarzeczny, U. (1999) *Economic Conditions of Tourism in Poland in the 1990s, Selected Issues of Tourism.* WSZ, Wrocław.

5 The role of foreign direct investment in the development of tourism in post-communist Hungary

Zsuzsanna Behringer and Kornélia Kiss

In the early 1990s Hungary was the world's fifth most-visited tourist destination country, but reductions in the number of curiosity-seeking tourists, and recurring conflicts in the Balkans, affected tourism in Hungary adversely: the country ranked 12th in 2001 and 13th in 2002. Tourism has been a major sector of Hungary's economy since the 1970s: its GDP contribution rate is significant, its role in foreign currency earnings is important and its ripple effect in creating jobs and boosting demand in many sectors of the economy is sizable. Although the late 1970s and early 1980s brought large-scale international development projects, tourism, especially the provision of accommodation and catering and travel agency services, became a very attractive target for foreign investors, particularly after 1989. Privatization boosted the inflow of foreign direct investment (FDI) mainly from the UK, Switzerland, France and Germany. Further, the late 1990s witnessed a new surge of interest from foreign investors that fuelled the current hotel construction boom. This chapter describes the most important tourism FDI projects implemented in the past decade, or currently in progress, in Hungary.

The Recent Development of Tourism in Hungary

Tourism in Hungary revealed five major trends during the 1990s (Behringer *et al.*, 2001). First, the number of foreign arrivals in Hungary dropped by 15.7% between 1990 and 2000, alongside a fivefold increase in the income from tourism during the same period, whilst average spending per visitor and trip increased almost sixfold. The decline in the number of arrivals in the past decade is attributable mostly to the reduction of shopping-motivated trips and retail trading by tourists from Central and Eastern European countries.

Second, there was a clear shift towards higher quality services in the use of commercial accommodation establishments. The proportion of guest-nights spent in hotels continued to rise, and the rate of increase in the number of guests staying at 4- or 5-star hotels especially so. The turnover of guests at commercial accommodation sites shows strong geographical and seasonal concentration: the national share of Budapest and Lake Balaton in total guest-nights spent in hotels and at other commercial accommodation in 2002 was 30% and 27%, respectively. The share of hotels in the occupation of commercial accommodation was 48% in Budapest and 19% at Lake Balaton.

Third, the purpose of a visit to Hungary includes vacationing and relaxation first of all, with business arrivals representing an increasingly pronounced share. Although a large proportion of arrivals from neighbouring countries came to Hungary on shopping trips in the early 1990s, the proportion of such visits continues to decline.

A fourth factor is the increasing role of domestic tourism since 1997. Numbers of domestic guest-nights grew by 4.6% per annum between 1997 and 2002 in commercial accommodation.

Finally, half of all international arrivals are day-trippers or hikers, and the proportion of such visits from Austria is close to 60%. Only about 12% of all international visitors stay in Hungary for more than a week.

Tourism Foreign Direct Investment in Hungary: Methodological Issues

The analytical method known as 'satellite accounting of tourism' has been developed for the purposes of quantifying its economic effects. The satellite account is a statistical and forecasting database that emphasizes the details and relationships typical of tourism, but is adjusted to the framework of traditional macro-economic analysis, i.e. to the system of national accounts. Only a very few countries have a satellite account of tourism at present, as developing the account and continuous data entry are extremely costly and labour-intensive. Such a satellite account that would allow the quantification of the economic effects of tourism is not currently available in Hungary, which explains the lack of exact information on the scope and size of tourism-related foreign direct investments.

At present, it is the Hungarian Central Statistical Office (CSO) that collects data regarding capital expenditure projects in tourism financed by foreign capital. The publications of the CSO, however, contain only information regarding capital expenditure projects in accommodation services and catering (which is a consolidated category), and that is only a fraction of all the investments implemented in the tourism sector. The following sections are devoted to analysing these FDI projects based on data supplied by the CSO (2002).

FDI Projects in the Hotel Sector in Hungary

Most of the country's hotels and the majority of catering establishments were destroyed during World War II. Post-war reconstruction employed several former hotel units for different purposes: they were transferred to government bodies or companies or were used as part of welfare tourism. New hotels were mainly constructed in industrial centres in the 1950s, ignoring the wider aspects of tourism. In Budapest, Hotel Gellért, which was constructed in the 18th century, and the Grand Hotel on Margaret Island were complemented by the Hungarian debut of two international hotel chains in the 1970s – Hilton and Intercontinental.

An agreement was signed with Austria between 1978 and 1985 for the deployment of a US$300 million credit facility earmarked for raising tourism product and service quality in Hungary. The credit was instrumental in expanding hotel capacity, mainly in the 4- and 5-star categories, by almost 7500 beds. Despite its disadvantages (Harbula, 1992), the credit arrangement permitted capital expenditure projects of this order of magnitude in Hungary for the first time. Each of the hotels constructed at the time was owned by the state, but the government had cooperation agreements in place with a variety of international hotel chains for trade-name use. Besides the construction of new hotels, the credit was also used to add new services (e.g. swimming pools, tennis courts) to a few hotels in the rest of the country. As a consequence, the 1970s can be seen as the first domestic boom period of capital expenditure for hotel projects, as the total domestic capacity of hotel beds increased by 45%.

Conclusion of the credit facility in 1985 decelerated capital expenditure substantially. The phasing out of government projects and the growing indebtedness of large hotel operations, due to the terms of the Austrian credit facility, saw construction projects limited to hotels, boarding houses and campsites funded privately or by new partnerships. The new legal environment that evolved in the late 1980s motivated a few domestic hotel chains to experiment with new arrangements for financing development projects.

Intending to raise funds for development and improvement, the domestic hotel chain, HungarHotels, made an abortive attempt at privatization in 1989, prior to political and economic change. The formation of the State

Property Agency (SPA) and the enforcement of the Act on Protecting National Assets soon marked the end of a period of spontaneous privatization.

The first programme of privatization was announced in September 1990 following the first free general elections. The first package of companies selected for privatization included state-owned companies with higher than average performance, in the hope that they would be sold more easily to private investors. That initial package featured: IBUSZ, a travel agency; Hungexpo, a fair operator and advertising firm; Centrum, a department store chain; and three large domestic hotel chains – HungarHotels, Pannónia and the Danubius Hotels group. After an invitation to tender by the SPA, 18 international consulting firms assisted these companies to succeed in becoming privatized despite the large amount of debt on their books. These consulting firms set the scene for privatization and prepared the target companies for the transactions. They had been selected by December 1990, and the first proposals were made in June of the following year.

Dismantling state ownership was laden with difficulties because it was hard to redesign the system of management and accounting in what used to be large socialist companies to the system of financing and accounting in developed market economies, which explains the need for involving international auditing companies. Tourism as a sector, however, enjoyed certain benefits, as the operators of accommodation sites and catering establishments had become familiar with the standard terms used in a market economy, due to the presence in Hungary of international companies (e.g. Hilton, Intercontinental, Hyatt, Forum) under trade-name licence agreements concluded earlier (Árva, 1998).

Of the companies selected for privatization, IBUSZ was the first to be floated in a public share offer, following recommendations from the consulting firm E. Rothschild, London. Initially, IBUSZ fared spectacularly for a short time, but the shares subsided and later on its market price fell below the issue price. Nomura, the advisor assisting Danubius, also recommended that the company should go public, but could not settle the Austrian debt

burden of Danubius, which prejudiced the chances of listing in the main list as the low level of expected gains was not particularly attractive for investors. Subsequently, Danubius was floated on the exchange through a programme designed for retail investors. Swissbank Corporation proposed the same to HungarHotels, but this transaction also failed to occur due to an extremely high level of corporate indebtedness. The strategy developed by Pannónia and J.P. Morgan was somewhat different in that they envisaged exchange flotation as the second step. The strategy devised by J.P. Morgan called for rounding up strategic investors for Pannónia, following legal transformation into a joint-stock company. After finding the appropriate strategic investor and strengthening the company through investments of the right size and the transfer of technology, a restructured and financially sound business could be listed on an exchange.

The privatization of HungarHotels was a long process involving several invitations to tender. First, Quintus, a Danish–Swedish real estate company, offered US$90 million for half of the shares. Yet the sale did not materialize in that stage of the spontaneous privatization, because the consortium failed to present bank guarantees and plans for long-term development. Swissbank Corporation, the advisor of HungarHotels, recommended following the example set by the Danubius hotel chain to have the company listed on an exchange. That strategy failed in 1991 as the indebtedness of the company deepened when the interest rate of its Austrian loan was raised. After several attempts, 20 hotels and five catering units were assigned to a newly formed joint-stock company in late 1991. The concept called for selling Hotel Royal and Duna Intercontinental separately. The US-owned Marriott hotel chain purchased Hotel Intercontinental for US$35 million in 1993. Hotel Forum was also divested from the HungarHotels chain and a separate company was set up to manage it in 1996. That left the chain with 14 hotels and the right to lease a restaurant, representing about US$85–90 million in registered capital, which formed the basis of the new privatization tender. In response to that call, Danubius submitted the highest bid and acquired 85% of HungarHotels. When the process of privatizing

Hotel Forum closed in the autumn of 1996, it was announced that Intercontinental Corporation had won the tender and the right to purchase 94.91% of the shares for US$49.4 million. The new shareholder soon spent US$10 million on refurbishing the hotel.

Pannónia Hotels was incorporated as a joint-stock company in December 1991. The local government held 18.5% of the shares and the State Property Agency was in charge of the remainder. HOPLA, a company affiliated to the French Accor group, submitted the best bid to purchase Pannónia – for US$57 million – in the privatization process and acquired 51% of the hotel chain on 30 November 1993. Later, the stake held by the French company reached 73%. The hotel chain comprises the 3-star hotels Volga and Aero, and the 4-star Korona, Buda-Penta and Novotel. The Pannónia chain has a single 5-star flagship, the Atrium Hyatt, which the Accor group wished to rename as Sofitel (a name used in the group to describe top-category hotel units). At that point the national airline Malév, the minority shareholder of Atrium Hyatt, blocked the process of renaming. In 2001, 50% of the Malév Hyatt Regency Hotel was bought by HVB Hungary. In the following year HVB bought the remaining 50%, and signed a 22-year leasing contract with the former owner of the hotel, Pannónia Hotels, to operate it.

Danubius was listed in 1992, and was fully privatized by 1994, with an ownership structure in place by 1996 whereby foreign shareholders held 55%, Hungarian retail investors 29%, 6.75% belonged to employees, broker firms held 6.22%, and an insurance fund owned 3.03%. The Danubius chain owns Budapest Hilton, Hotel Termál and Grand Hotel on Margaret Island, Hotel Gellért, Hotel Termál and Hotel Aqua in Hévíz, Hotel Termál in Sárvár, Hotel Termál and Sporthotel in Bük and Hotel Helikon in Keszthely. The core business of the hotel chain involves the operation of curative centre hotels, whose clientèle is mainly international. On the other hand, domestic tourists predominate in the units of Béta Hotels, a fully-owned subsidiary of Danubius. The Béta chain has 13 hotels, mostly in the countryside. Hotel Hélia joined the chain in 1997, and the Danubius group followed international trends by splitting into two

on 1 January 1999. The two joint-stock companies created through the de-merger are responsible for managing the units and operations and are still owned by Danubius. The hotel chain neither constructs nor opens new units due to the austerity of financial regulations in Hungary; it regards the refurbishment of its existing member hotels as its main objective.

With the first privatization programme over, a second was announced in 1995 and the package offered for sale to private investors included two additional hotel chains: Eravis and Hunguest. The legal predecessor of Eravis had started using the former dormitories where state-owned companies used to house holiday-making workers and their families. Starting in 1993, the core business of Eravis involved the operation of 1- and 2-star hotels, employee hotels and student hostels. As an operator of mostly low-category accommodation units, Eravis was successfully listed on a stock exchange in the early 1990s and private investors acquired majority ownership in the company. ÁPV, the property holding of the state, announced the privatization of the hotel chain in November 1995 and commissioned the brokerage, K&H, to act as advisor and lead manager. HB Westminster II acquired a majority stake in Eravis in mid-February 1996, but the new shareholder did not replace the management of Eravis. The company eventually established a strategic alliance with Hunguest: 50% of Hunguest, the former socialist recreation company, was transferred to Eravis, which came to hold most of the medium-category hotels in the countryside. Since that time, the two companies have been engaged in similar lines of business and operate units of a similar category. As a member of the consortium, Eravis helps Hunguest operate, modernize and refurbish 25 hotels outside Budapest.

Hunguest was established in 1993 as the operator of 36 property units of the National Recreation Fund, the entity responsible for managing trade union-financed vacations. Domestic and foreign tourists spent two million guest-nights in these hotels in 1995. More than half of the guest nights were supported financially by the Fund, as the prime responsibility of the hotel group involves the long-term provision of domestic holidays at subsidized rates.

However, the proportion of foreign guests using these hotels has been increasing. One of the subsidiaries of Hunguest, Hunguest Hotels, is the entity responsible for operating hotels in the group. For the refurbishment and modernization of its hotels, the strategic agreement for the purchase of 50% of Hunguest by Eravis was concluded in December 1996. The refurbishment projects are funded from two sources: the earnings of the company and the capital leveraged by Magyar Beruházási és Fejlesztési Bank.

The second boom in hotel construction started in 1999 as a result both of granting favourable taxation terms to foreign investors of capital in Hungary and the stable political,

economic and financial environment of the country. That boom increased the number of 5-star hotels from seven to 11 in 2000, including nine units in the capital. The investment value of the 5-star hotels opened in 2000 is close to US$100 million, whilst that of the 4-star is US$70 million. The developments are concentrated mainly in the capital, but the provinces, for instance Hévíz, Szeged and Tokaj, also have six new 4-star hotels and a 5-star unit.

Three 5-star hotels were refurbished and reopened in 2002, adding 735 new rooms to the existing supply. Of the 4-star hotels, five units were to be rebuilt to add 931 new rooms to the capacity existing in this category.

Table 5.1. Hungary: the features and investment value of 4- and 5-star hotels commissioned in 2000.

Name	Category	Number of rooms	Number of beds	Investment value (US$ million)
Art'otel	4*	166	322	10
Burg Hotel Budapest	4*	28	56	na
BWD Lido Hotel és Sportc.	4*	57	100	3
Hotel Le Meridien	5*	218	452	45
Hotel Páva Plaza	4*	206	410	15
Mercure Metropol	4*	130	230	na
Mercure Múzeum	4*	54	109	na
Starlight Suiten	4*	55	126	15
Sydney Apartment Hotel II	5*	97	230	27
WestEnd Hilton	4*	230	470	24
Total	–	1241	2505	139

na, data not available.
Source: Deputy State Secretariat Responsible for Tourism, Ministry of Economic Affairs (unpublished).

Table 5.2. Hungary: the features and investment value of 4- and 5-star hotels commissioned in 2001 and 2002.

Name	Category	Number of rooms	Investment value (US$ million)
Alexandra	4*	21	3
Andrássy Hotel	4*	72	10
Corinthia Royal	5*	400	89
Crowne Plaza	4*	310	75
Four Seasons	5*	155	85
Holiday Inn Rózsadomb	4*	301	35
Novotel Palace Budapest	4*	227	na
Total	–	1486	297

na, data not available.
Source: Deputy State Secretariat Responsible for Tourism, Ministry of Economic Affairs (unpublished).

Despite the high number of completed capital expenditure projects, the willingness to invest is still strong. One of the latest plans is the Millennium City Centre. The centre is designed to occupy the waterfront between Petőfi Bridge and Lágymányosi Bridge, and is reckoned to be the largest hotel investment project of the decade. The development plans envisage the construction of five new hotels and a spectacular promenade to be called the 'Avenue of Restaurants'. The Millennium City Centre would feature hotels with 420, 355, 320 and 180 rooms, including a single 3-star unit, with each of the other hotels classified as 4- or 5-star establishments. If the project is implemented, 1620 rooms which are due to be opened to the public in 2004/05 will be added to the existing capacity of accommodation in Budapest city centre. Accor-Pannónia is planning the project of the largest portion of this building complex: the group intends to construct three units, including a 5-star Sofitel hotel, in this new part of the city.

The Italian-owned Boscolo Group hallmarks the second major hotel construction project. The group won the tender invited for the New York Palace, considered to be the gem of the stock of buildings in Budapest, which has been closed for a long time and is in a state of deterioration. The Group wishes to operate it as a luxury hotel after complete reconstruction.

Castle renovation, and the construction of health spas, fitness centres, theme parks and conference centres are becoming more and more attractive investment targets, but several foreign investors seem to continue to prefer investing in hotels. Hotel investments are reckoned to play a dominant role among tourism projects because the size of investment opportunities in Hungary is not attractive for foreign investors, disregarding complex product development, theme parks and similar projects.

Typical Features of Accommodation and Catering Projects Financed by Foreign Direct Investment and Foreign Companies

The value of foreign direct investment in the provision of accommodation and catering subdivisions of tourism in Hungary is by far the largest among the former soviet satellite states in Central and Eastern Europe. The growth rate of the volume of capital invested directly by foreigners in accommodation and catering projects is much faster than the rate of growth of the total volume of capital expenditure.

The sudden surge in the number of businesses was the most spectacular change within the tourism sector in the early 1990s. However, statistical data regarding foreign ownership of businesses by industrial sector are only available from 1993, hence it is impossible to demonstrate an increase in the number of foreign enterprises that would correspond to the large increase in the number of businesses in the early 1990s. By 1993, the marked propensity to set up businesses had slackened somewhat and the number of enterprises held by foreigners increased only by a factor of 1.3 between 1993 and 2001.

The rate of growth of the registered capital invested by the foreign providers of accommodation and catering establishments was only slightly higher than the rate of growth in the number of businesses in the same period, as registered capital grew by a factor of 1.62 between 1993 and 1996. As far as the ratio of foreign funds in registered capital is concerned, the tendency of continuous increase in the foreign portion is worth noting: except for the first 3 years in the reviewed period, the ratio of foreign funds continued to climb, to reach 87% by 2001. Between 1995 and 2001 the registered capital of businesses held by foreigners grew at the rate described above, but the ratio of foreign capital to registered capital total declined modestly.

An analysis of foreign businesses between 1993 and 1999, by the size of the share held by foreigners, reveals the same increasing preference of foreign investors for exclusive foreign ownership or foreign majority holdings in the accommodation and catering sectors.

Foreign investors tended to establish fully owned operations in most of the cases in the early 1990s. Nevertheless, enterprises with a foreign majority predominate on the basis of the foreign share in registered capital. By the end of the decade the composition had changed radically, in that most businesses with foreign interests were fully owned by foreigners and fully owned foreign firms also had the largest share in the registered capital of businesses held by foreigners.

Accommodation providers and catering establishments with foreign participation represented almost 4.5% of all the enterprises existing in 1994. That ratio had increased to 4.8%, after minor fluctuations, by 2001. The registered capital of businesses with foreign participation fell short of the proportion measured by the number of businesses with foreign participation, and was on the decline, from 3.96% in 1994 to 1.38% in 2001.

The Composition of Foreign Direct Investment in Tourism by Source Country

Information regarding the composition of foreign direct investment in tourism by source country is limited to opinions formed by professionals and estimates that identify investors from Austria, Switzerland, France (e.g. Accor-Pannónia), the UK (e.g. Danubius), the USA, Germany and Israel (especially in hotels) as the major sources of funds. The Central Statistical Office produces information based on data reported by companies with the largest amount of foreign capital, hence its statistical yearbooks publish data for the countries that play the greatest role in the total flow of foreign direct investments into Hungary, i.e. data for Germany, Austria, Switzerland, the Netherlands, Belgium, France, Italy, the UK, the USA and Japan. The following section is devoted to describing the flow of FDI from these countries for the provision of accommodation and catering between 1993 and 2001.

In 2001, the largest proportion of FDI for the provision of accommodation and catering in Hungary originated from the UK (Table 15.3), although the popularity of investments by businesses with UK participation had fluctuated in the previous decade. The equity of businesses with 100% British participation came to HUF (Hungarian *forints*) 15.7 billion in 2001. More than 21% of the total FDI from the UK was in the accommodation and catering sectors. The second largest foreign direct investor was the USA: 14% of the accommodation and catering investments came from there. The equity of businesses with US partici-

pation was the largest amount, almost HUF 36 billion, but the actual American portion was low, at HUF 10.2 billion.

France and Austria were the third and fourth largest providers of FDI for the accommodation and catering sectors. Whilst the registered capital of businesses with Austrian participation and the size of Austrian FDI flow have continued to increase in recent years, the amount of Austrian capital flowing into the provision of services and catering has fluctuated widely. The Netherlands and Switzerland were also important investors in the accommodation and catering sectors in 2001, with 3.2 and 2.7% respectively of the total accommodation and catering investments.

German and Italian investors are important for Hungary in terms of the total amount of foreign direct investment. An analysis of their investments in accommodation and catering businesses, however, reveals a more subdued role as their annual inflow of funds is typically below, or at most, HUF 1.0 billion. Businesses with German participation in the provision of accommodation and catering invested around HUF 3.2 billion worth of capital in 1993, but subsequently German investors appear to have lost interest in these sectors, with investment declining from 1994 to a level of just HUF 0.1 billion by 2001, representing no more than 0.01% of the total FDI from Germany in Hungary.

Conclusion: the Future of Foreign Direct Investments in Tourism in Hungary

A strengthening of the Hungarian economy and that of neighbouring countries, coupled to EU accession, may create more favourable conditions for foreign direct investment inflows into Hungary. The restructuring of foreign visitor markets and the strengthening of domestic tourism, coupled with the increasing proportion of tourists seeking quality services, may suggest a favourable environment for foreign investors. The largest international tourism associations, including the WTO and the European Travel Commission (ETC, 2002), identify Central Europe as one of the tourist destinations with the largest potential for dynamic growth.

Table 5.3. Hungary: distribution of FDI flows for the provision of accommodation and catering, 2001.

Country	Equity of business with foreign participation (HUF billion)		Including provision of accommodation and catering (HUF billion)		% share of provision of accommodation and catering	% share of FDI for the provision of accommodation and catering
	Equity	Amount sourced from country itself	Equity	Amount sourced from country itself		
UK	125.6	83.6	15.7	15.7	18.8	21.6
USA	607.4	557.2	35.8	10.2	1.8	14.0
France	456.3	341.9	7.6	7.5	2.2	10.3
Austria	869.8	689.6	4.6	4.2	0.6	5.8
Netherlands	995.2	944.4	2.3	2.3	0.2	3.2
Switzerland	109.3	89.6	2.1	2.0	2.2	2.7
Italy	140.5	127.0	0.2	0.2	0.2	0.3
Germany	2571.2	2104.7	0.1	0.1	0.0	0.1
Other	1512.9	1257.2	33.4	30.6	2.4	42.0
Total	7388.2	6195.2	101.8	72.8	1.2	100.0

Source: CSO, 2002.

Hungary's GDP has been increasing steadily since the mid-1990s (see Table 1.1). Economists forecast continued GDP growth for a few years to come, and that the CEECs and especially Hungary will out-perform the member states of the European Union in terms of GDP growth. Several neighbours in CEE are already among Hungary's most important source markets, although at present most of them only send hikers, and their spending on the services offered by tourism is low. The current and expected increase in the standard of living in these countries will strengthen outbound tourism in the years to come, and Hungary will appear as one of the possible destinations. Seen from the perspective of foreign direct investments into tourism, the continuous improvement of performance of the national economy in Hungary may be important for two reasons. On the one hand, a stable and reliable economy constitutes an attractive investment environment for foreigners. On the other hand, income growth creates improved living conditions for local inhabitants who then spend larger amounts of their higher disposable income on recreation and travelling.

The decline in numbers of foreign visitors to Hungary by 15.7% between 1990 and 2001 (WTO, 2002a,b), and the simultaneous threefold increase to almost €3.7 billion in 2002 in receipts from foreign tourism, represent a positive opportunity that emerges from the current domestic environment. The appeal to foreign visitors is being restructured in a way that encourages higher-value tourist demand. Another opportunity relates to the growing recognition of the role of tourism in the national economy. Currently, tourism ranks high among the sectors prioritized for support by the government.

The increase in leisure time and disposable income is an international trend in the developed part of the world, and the number and intensity of trips also keeps growing. Hungary should be able to become a suitable destination for spending a second or third vacation. The country's cultural values and traditions are rich, thermal water reserves are significant and spa resorts well developed with a long-standing international reputation for quality. These attractions should generate permanent demand for Hungary as a destination for high-income tourists from both traditional markets and new source countries.

References

Árva, L. (1998) *Diczházi Bertalan: Globalizáció és Külföldi Tőkeberuházások Magyarországon.* Kairosz Kiadó, Budapest.

Behringer, Z., Király, G., Kiss, K., Rátz, T. and Török, P. (2001) *Idegenforgalmi és Vendéglátói Ismeretek* I, II. Szókratész Kiadó, Budapest.

CSO (Central Statistical Office) (2002) *Központi Statisztikai Hivatal Statisztikai Évkönyv.* CSO, Budapest.

ETC (European Travel Commission) (2002) *ETC Annual Report 2001.* ETC, Brussels <http://www.etc-europe-travel.org/Rapp2001.pdf>.

Harbula, G. (1992) A Magyar Szállodaipar Privatizációja. Unpublished PhD thesis, University of Economic Sciences, Budapest.

WTO (World Tourism Organization) (2002a) *Compendium of Tourism Statistics 1996–2000.* WTO, Madrid.

WTO (World Tourism Organization) (2002b) *Tourism Highlights 2002.* WTO, Madrid <http://www.world-tourism.org>.

6 Estonian national heritage, tourism, and paradoxes of transformation

Barry Worthington

Introduction

The study of heritage tourism enshrines a very curious paradox. On the one hand, it is analysed from a variety of academic perspectives. We may adopt a post-modernist approach and see it as a 'social construct' (Urry, 2001), and concern ourselves with its 'commodification'. On the other hand, following in the path of Robert Hewison (1987), we can also raise issues relating to 'authenticity' and postulate the concept of a 'heritage industry'. Whatever our approach, we engage in the dissection of a specimen, placing 'heritage' in a cultural or sociological context, and touching upon issues of personal and communal identity in the process.

Having made our analysis, we often pause and go no further. We may not attempt to connect our academic interest with other areas of concern. This is a pity, for there are several instances where 'heritage' is not purely a saleable commodity or an interesting cultural phenomenon, but a live and volatile issue. We may be perfectly happy to incorporate 'heritage' into the tourism product, but shy away from the possible function that such 'heritage' may have in its relationship to nationalism and political identity. There are few studies exploring this relationship. Palmer has highlighted the function of tourism in the construction of an English identity, but he is more interested in the implications of this interaction for the man-

agement and promotion of heritage sites (Palmer, 1999). Presumably, the actual implications for the evolution of the English cultural identity do not present any significant issues. Conversely, studies of the evolution of Estonian nationalism are legion, but nearly all are written from the point of view of the political scientist and, sometimes, by the political geographer (Berg and Oras, 2000). Consequently, this chapter employs Estonia as a case study in examining the living relationship between national heritage and tourism.

Context

Estonia, the furthest north of the three former Soviet Baltic States (Fig. 6.1), regained its political independence in the *annus mirabilis* of 1991. However, the process of political emancipation was complicated by another process – the transition from a Soviet-dominated economy and society. In their 'singing revolution', Estonians' culture constituted an important weapon – a catalyst for political change. As 'heritage' is such a major plank in the Estonian tourism product, the question is raised: what was the relationship between national heritage and Estonian heritage tourism?

There were two overlapping but distinct Estonian nationalist elements throughout the *perestroika* era of the second half of the 1980s

Fig. 6.1. Estonia.

Fig. 6.2. Estonia: the Old Town, Tallinn.

(Lieven, 1993). As Estonia was a Soviet republic, the local communist party, the *komsomol* (party youth wing), and the republic's *soviet* (governing assembly) had some degree of autonomy from Moscow. The advent of a new political generation could use this situation to loosen the ties with the centre and introduce limited economic and social reforms. Thus, the IME project was created, a national administrative strategy designed to obtain a wide degree

Fig. 6.3. Estonia: Tallinn Town Hall; the square is now car-free.

path to economic autonomy and reform; Trivime Velliste, the first EHS president, advocated the so-called 'Hungarian Path' (economic reform within the existing communist system) for Estonia as a 'socialist' country outside the USSR. Nevertheless, the EHS distrusted the existing institutions of government, considering them tainted with Soviet practice. It did not participate in the first Supreme Soviet elections, although it did not advocate a boycott (Taagepera, 1993).

The EHS developed into a major cultural movement. The re-dedication of Estonian war memorials, some new, some re-erected from dismantled parts hidden in 1940, began in June 1988, with ceremonies in which clergy, church choirs and reconstituted scout units participated. The EHS was active in re-introducing the use of the national flag in the spring of 1988. However, the most ambitious project was the creation of citizens' committees to produce an alternative framework of civil society upon which political independence could be based (Taagepera, 1993).

These committees emphasized the continuity of pre-war Estonian institutions and registered citizens on the basis of pre-war citizenship or descent from pre-war citizens. By September 1989 over 300,000 had registered, including numbers of post-war immigrants applying for citizenship. About one tenth of the total may have been of Russian or of foreign origin. The EHS had Russian and Jewish branches at this time, to accommodate non-Estonian sympathizers (M. Aru, 1995, personal communication). The spread of these committees was slower than anticipated since policy dictated that local people, not emissaries from Tallinn, should be the progenitors. At length, the Society participated in the formation of a General Committee of Citizens of the Republic of Estonia on 11 November 1989, which was opposed to the Popular Front's acceptance of the practical need to work through existing Soviet institutions. The Heritage Society took 104 seats in the ensuing elections to the Estonian Congress on 24 February 1990 (Taagepera, 1993).

However, events overtook the political development of the Estonian Heritage Society. The circumstances of the abortive August coup in Moscow resulted in a joint decision with the

of economic autonomy from Moscow. This movement, working through existing Soviet institutions, was known as the Popular Front. However, in a parallel development, another widespread movement appeared with the objective of creating an 'alternative' society to challenge the Soviet model. It focused its attention upon re-creating historical and cultural traditions through revived civil institutions.

The Estonian Heritage Society (EHS) furnished a good illustration of the process of reclaiming the civil sphere, thus setting the context for important tourism developments. Its origins lay in a number of cultural heritage protection clubs in the Tallinn area, and it was founded in that city in December 1987 (Ruutsoo, 1993). Such *ad hoc* groups had concerned themselves with the preservation and restoration of national monuments, starting with churches and neglected churchyards or cemeteries. The working parties included academics and students (O. Sandrak, 1992, personal communication).

Such activities were, by their nature, political. 'Heritage' became a radical issue and the movement found it necessary to collaborate with the Popular Front. It approved the official

Popular Front to declare immediate independence – a declaration recognized by Boris Yeltsin, now assuming the reins of power in Moscow. Thus different streams of independence merged. An immediate effect was the emasculation of the heritage movement. Upon independence being achieved, many thought that the Society's task had been accomplished. Many EHS members went on to develop a separate political career and left the rump of the organization to its own devices.

The remainder of the original members stayed to continue what they saw as 'patriotic activity', promoting historical research and publications about the Soviet occupation and the fate of Estonian deportees earlier in the century. The definition of 'heritage' extended to architecture and artefacts. By 1995, the EHS board was concentrating upon the preservation of 'cultural heritage'. The then chairman believed that all periods of Estonian history and its structures, including Soviet buildings and war memorials, were worthy of conservation. The board was aware of the importance of industrial heritage and industrial archaeology, and had already been involved in disputes with Tallinn city council concerning proposals for the dockland area's development. These activities took place in what must have been a very new and alien context: the once-important Estonian Heritage Society now operated as one lobby group amongst many, all competing for the attention of government ministers (M. Aru, 1995, personal communication).

The Heritage Society's headquarters in Tallinn provided facilities and a home for associated groups promoting other areas of heritage conservation. Among these was one concerned with the development of a railway museum at Lavasarre, near Pärnu, and another group promoting the restoration of a railway as a tourist attraction on the island of Naissar (M. Helme, 1995, personal communication). The main concern of the Heritage Society chairman appeared to be the conservation of the traditional wooden houses in the suburbs of Tallinn and elsewhere. He looked to the possibility of the introduction of communal laws – on the Swedish model of town planning – and the establishment of conservation areas for these traditional suburbs.

Had the EHS maintained its prestige and influence, it surely would have proved a major influence on the post-Soviet Estonian tourism industry. After all, 'heritage' was such an important part of the Estonian tourist product, and had played a key role in the process of national emancipation, yet this assumption proved to be false. The Society did not harbour any particular views on the subject; they were aware that possible economic and cultural dangers relating to the impact of mass or uncontrolled tourism might exist, but the chairman thought that there were very few areas in Estonia where tourist access should be limited. This was in spite of the fact that Helsinki ferry traffic was approaching 1.4 million passengers per annum (the total population of greater Tallinn being less than half a million).

Indeed, there appeared to be little awareness of the problem of the irregular distribution of tourism flows in Estonia. There was some discussion about the desirability of a visitor tax, but only from the point of view of raising revenue for conservation measures, not as a means of controlling visitor numbers in particular areas. It seemed clear that debates elsewhere about tourism impacts and the management of tourism flows, with their implications for the future of Estonian heritage and national identity, did not constitute an issue (M. Aru, 1995, personal communication).

In its new incarnation, the Estonian Heritage Society was handicapped by its ignorance of current heritage, conservation and tourism issues, even extending to issues concerning the nature of environmental interpretation for the foreign visitor. Lack of resources and lack of knowledge was emphasized by the then leadership. Consequently, it appeared that the influence of this grass-roots nationalist organization upon Estonian tourism was negligible. Whilst it is true that events overtook a slowly developing process of cultural awareness, lack of adequate knowledge of the nature of Western tourism, together with its cultural interests and sympathies exemplified in heritage tourism, was a crucial factor. A great opportunity was surely missed.

Perceptions of Heritage

If the grass-roots heritage movement had little impact on the evolution of the Estonian tourism product, how did Estonians involved in the tourism industry perceive it at the time independence was regained? Did national feeling play any part in this? Two surveys of Estonian tourist literature were carried out during this period by the author. The first concerned tour company brochures collected in November 1993, which reflected the 2-year period immediately after independence. This survey focused on tourist destinations indicated in the publicity, and was therefore concerned with Estonian perceptions of the perceived strengths of their tourism product. Summary results are indicated in Table 6.1.

The popularity of Tallinn, given its pre-war status and Soviet-era significance as a tourist destination and arrival point, comes as no surprise. Indeed, references to the capital dominated the tourist literature in the sample, in the form of promotions for coach or walking tours and excursions, including attractions in the environs such as Pirita (with its ruined monastery, historic ships and beach) and the folk museum at Rocca. The usual itinerary for the capital extended to 2 or 3 days' duration, although the most popular excursion was the day trip, catering for the growing ferry and cruise market.

The references neatly fell into two sections. Over half of them (represented by Tallinn, Tartu, Pärnu and Viljandi) corresponded to the former Intourist itinerary of the *perestroika* period (Soviet Baltic republics). Tallinn had been marketed as an international destination since 1965. Although the rest of these locations may not have been accessible to foreign tourists until the eve of independence, they reflected the tourism product fashioned by the trade union-based Tourism Councils for the rapidly growing Soviet domestic market.

Soviet cultural tourism traditions were essentially the same. They were commodifications of a mainly urban heritage, representing a 'history' and 'culture' in the form of historic sites, architecture, and a variety of physical artefacts. This form of cultural consumption, of course, lent itself to interpretation in a particular ideological way. Even the folk heritage, for which Estonia is so famous, was presented in carefully packaged performances to ensure a politically neutral message to the audience. This form of passive consumption of a 'packaged culture', defused

Table 6.1. Estonian destinations featured in tour company literature, 1991–1992.

Destination	Number of mentions in brochures
Tallinn (day tours)	10
Tallinn (more than 1 day)	4
Saaremaa (Estonian island destination)	9
Tartu (historic city)	8
Pärnu (seaside resort)	8
Lahemaa National Park	7
Haapsalu (seaside resort)	6
Narva (border town)	5
Viljandi (historic town)	3
Hiiuma (Estonian island)	2
Rahkvere (castle)	2
Pyhitsi Convent (Kuramae)	2
Polva (country town)	2

Source: author's original research.

of any possible nationalist or social issues, dominated Estonian tourism. Other activities, such as sea bathing, were seen purely as 'individual recreation'. The countryside or nature tourism was not featured.

The remaining references were dominated by Saaremaa, the Lahemaa National Park and Haapsalu – all new tourist destinations. If the previous destinations suggested a slavish adherence to Soviet practice, surely, here was a token of the future of Estonian tourism, the promotion of the Estonian cultural heartland? Indeed, the island of Saaremaa, featured by nine companies, was the second most popular destination proposed to the visitor. It had previously been a restricted area, owing to the presence of Soviet military bases, but that alone did not explain the popularity of this new location. The island was the most accessible of the Estonian archipelago, by means of a short ferry crossing from Virtsu. The castle at Kuresaare was famed as the best preserved in the country, but perhaps the real importance of the island as a tourist destination originated with its associations in the Estonian mind with the concept of national identity. The location was packaged as the 'true Estonia', a unique selling proposition for the international market as an unspoilt haven of traditional Estonian life. Thus, 'as a result of their relative isolation, it is on these two islands (Saaremaa and Hiiuma) that true Estonian culture has been preserved. The islands are totally free from any sovietization, and, for that matter, of any of the newer encroaching westernization' (Baltic Tours, 1992).

Yet all was not as it seemed. This description probably owed a great deal of its inspiration to the Finnish market for which it was produced (Baltic Tours was a joint Finnish–Estonian enterprise), and Saaremaa had a very high market profile in Finland – attributed by some to the singing of the late Georg Ots (J. Raag, 1995, personal communication). Ots remains a popular musical icon in Finland – not surprisingly, the Helsinki ferry was named after him in Soviet times. Many of his songs are set on this Estonian island, and some of his music is still used for dancing – the 'Saaremaa Waltz' created its own publicity. Thus, though Saaremaa was a showcase of Estonian national pride, it was a showcase that was already very familiar to Estonia's largest foreign tourist market. The fact remained that the inclusion of this island in tourism itineraries was largely dictated by the demands of an emerging foreign market.

Hiiuma was featured by only two companies in comparison; possibly a reflection of the size and easier accessibility of Saaremaa, its main island competitor. The quieter and more rural setting of Hiiuma island would qualify it as a countryside attraction, but nature tourism remained an unappreciated element of Estonia's tourism product at this date. Indeed, the only true attraction of this kind to be promoted in the brochures was the Lahemaa National Park, on the mainland, and this single example of the Estonian countryside probably owed more to its convenient location on the Tallinn–Narva transit route, and its proximity to the capital, than to any particular quality of its environment. By contrast, the attractive but isolated Matsalu reserve in western Estonia, popular with Estonian birdwatchers, was ignored.

More attention was lavished upon nearby Haapsalu, a seaside town with an ancient castle, which was featured by six companies. Static concepts of heritage obviously took precedence over activities associated with domestic recreation in Soviet times. 'Heritage' was perceived as the obvious lodestar of the foreign visitor, and the importance of the countryside as an attraction was consequently downgraded. Spectacular scenery proved an exception to this rule, but Estonia obviously did not appear to qualify in this area. Clearly, Soviet attitudes still permeated the new tourism industry.

Narva appeared as a destination in five brochures, though this was largely due to its position as a transit point between Tallinn and St Petersburg. The majority Russian population in Narva ensures that the city could not be described as particularly 'Estonian' in character. The old town was largely destroyed in World War II, although the impressive (restored) castle remained as a feature.

The tourism product largely remained what it had been in Soviet times: focused centres of urban heritage with the addition of Saaremaa as a Finnish consumer-led development. Even Saaremaa was marketed in commodified heritage terms. Some anomalies from the Soviet era remained as the tour companies emerging from the trade union-dominated Tourism Council organization capitalized upon their former contacts. The Tallinn Travel Bureau, 'one of the oldest travel agencies in Estonia – thirty years experience,' advertised an 'exclusive' tour to Paldiski, described as 'a closed army residence for 54 years.' This still-restricted area, garrisoned by Russian troops, appears to have been marketed as an attraction through a lucrative arrangement with the military authorities, although the copy referred to 'the invasion of the Soviet army after the Molotov-Ribbentrop treaty in 1939'. The Tallinn Tour Agency suggested the dubious dockside district of Kopli as an interesting venue with visits to the fishing museum and devastated areas of a former Soviet military camp, and the company continued to offer visits to industrial enterprises and institutions in the Soviet manner. The old tourism attitudes died hard.

How did the new Estonian tourism industry see itself in relation to other Baltic tourism products? This could be demonstrated by placing the analysis of the tourism literature in a larger regional context. The results revealed a parochial attitude, which may have been a consequence of initial preoccupations with the Finnish tripper market. Only three companies dealt with St Petersburg, and a similar number offered 'Via Baltica' or 'Baltic Capital' tours. The Raetourist company was the most imaginative in this direction (E. Noorkov, 1995, personal communication). The most popular destination beyond Estonia was Riga, featured by seven companies, yet only one company included the Palace of Rundale, 'The Baltic Versailles', in their itinerary, even though it is only a short distance from the Latvian capital. Vilnius was featured by four companies, and Kaunas by three. Estonian companies offered the use of contacts in other Baltic states for the production of 'tailor-made' itineraries, but it was obvious that a 'Baltic identity' did not feature in the packaging of their products. They seemed content to service the needs of foreign operators who understood the international market perception of the Baltic states as one destination.

The results of this survey also suggested some interesting reflections concerning attitudes to special attractions. Two companies included Pyhitsi convent (Kuramae) in their programme. The only functioning Orthodox Convent in Estonia, Pyhitsi was the equivalent of Melk in Austria or Buckfast Abbey in Britain, and featured on the pre-war tourist circuit. Its neglect after independence in the early 1990s raised interesting questions about Estonian perceptions. Was this an example of cultural discrimination (against the Russian-speaking inmates) or a misunderstanding of the tastes of Western tourists?

Two companies included art and music in their brochures, but little interest was exhibited in any form of special interest or activity tourist experience beyond organizing a few packaged 'performances'. Others were willing to produce 'tailor-made' special interest packages, but a great gulf sometimes existed between client and provider, as activity or special interest holidays were not a particular feature of foreign tourism in Soviet Estonia. Thus an Estonian company greeted a request (at a major travel fair in London in November 1993) for a railway enthusiast holiday itinerary with blank amazement. 'But no-one would travel on our trains,' was the reply, 'they are too slow'. A request from a company specializing in tour packages for aviation enthusiasts met a similar degree of incomprehension. Hunting, fishing, folk music and health tourism were understood as legitimate interests, but not promoted in the tourist literature, the latter probably as a consequence of concern about the poor standards of accommodation at sanatoria.

The conclusion was that the Estonian tourism industry still approached 'heritage tourism' in a 'Soviet' way. It was seen as an essentially packaged form of passive consumption, an assembly of buildings, artefacts and scenery, politically neutral and involving little or no interaction with the host community. Examples of performance art, such as folk music, were similarly packaged and did not involve any of the indigenous song festivals. Whatever innovations under way were clearly consumer led, or instigated by foreign partners.

Change and Diversification

A second literature survey undertaken in 1995 found that brochures had achieved a degree of sophistication that enabled a more comprehensive analysis to be undertaken. The availability of the material did not permit an exact match as far as the contents of the two samples are concerned. Nevertheless, both samples remain representative of their respective periods, and demonstrate a degree of progression (Tables 6.2 and 6.3).

Both tables illustrate substantial change. The traditional locations featured very strongly, but the product had diversified into other areas. The north-east of the country had developed a tourist circuit embracing the castle at Rakhvere with six companies instead of two, and Pyhitsi Convent with nine companies when previously there had been two, and hinging on Narva with five companies now grown to eight. However, this circuit was still very much within the context of transit between Tallinn and St Petersburg.

The realization that Pyhitsi was a particular attraction to Western visitors was an interesting development. The possible attendance by tourists at the choral service and tours by the nuns had introduced a 'pro-active' element into the previous 'passive' consumption of the heritage package in this area.

Another significant development was the opening up of the previously neglected southern area of the country. References to Viljandi, with its picturesque castle and lake, increased from three to seven companies. The districts of Voru, Otopaa, Elva and Valga made their appearance on the tourist map. Heritage remained an important aspect of these areas, but the principal selling point of the south

Table 6.2. Estonian destinations featured in tour company literature, 1995–1996.

Destination		Number of mentions in brochures
Tallinn		12
Tallinn Kadriorg Palace		8
Tallinn Pirita	Environs	6
Tallinn Rocca (Folk museum)		8
Tartu		12 (4 including Polva)
Pärnu		11
Viljandi		7
Saaremaa		12
Hiiuma	Estonian islands	3
Haapsalu		8
Lahemaa National Park		10
Narva		8
Rahkvere (castle)	Russian	6 (3 including Padise Castle)
Pyhitsi Convent (Kuramae)	Transit	9
Pechory Convent (in Russia)	Route	6
Kihnu (island)	'Ethnic'	4
Vormsi (island)	islands	2
Otopaa		2
Kuttsemae (ski centre)		1
Elva	Southern	1
Valga	Estonia	1
Voru		4

Source: author's original research.

Table 6.3. Specialist package holidays in Estonia, 1995–1996.

Type of package	Number of companies
Russia with Estonia	9
Baltic tour	11
Hunting	4
Cultural/event promotion	5
Estonian folk performance	8
Churches and monasteries	3
Golf	3
Farm tourism	2
Soviet relics (e.g. Paldiski)	3

Source: author's original research.

appeared to be nature tourism and associated activities such as hiking and winter sports. Polva, with its lake and sandstone caves, was featured by four companies.

Some of the interest in this area had been the consequence of local initiatives, and the Voru Farmers Union was instrumental in promoting farm-based tourism in the district. Enterprise had been shown in linking the tourist itinerary with Pechory Convent, across the frontier in Russia, which was featured by six tour companies. This attraction, popular in pre-war times, involved tours of the monastery and catacombs conducted by the monks.

The new emphasis upon nature and living heritage appeared elsewhere in the survey. Ten companies featured Lahemaa National Park for the general tourist, but the most significant development lay in the extension of interest in the Estonian islands. Hiiuma remained limited in its appeal, a reflection, perhaps, of its awkward geographical location, as it is reached by a ferry from a remoter part of the Estonian coastline. Saaremaa, by contrast, is a short sail from a point near the busy Via Baltica. Indeed, Saaremaa maintained its leading position, being featured by 12 companies and was included in a tourist circuit from Tallinn to Pärnu.

A notable development, however, was the appearance of a new form of 'interactive' island tourism, involving beach barbecues and folk evenings. Kihnu, in Pärnu bay, appeared in the literature of four companies. The chief attraction was the survival of folk custom and costume on the island, and the destination has been popularized by the legend of 'Kihnu John,' one of Scandinavia's most famous seafarers. This form of 'living heritage' was beginning to be marketed by a trust acting on behalf of the community (Kihnuranzl, nd). The similar island of Vormsi was included in two brochures.

Another new feature revealed by the survey was the appearance of the specialist tourism packages. Some of these were quite imaginative, and included a 'retreat' at the Dominican monastery at Tallinn with appropriate music. The majority of these offerings were based upon cultural tours and events; folk performances were included by eight companies. Nevertheless, although the importance of an active form of heritage as part of the tourism product now seemed to be understood in some parts of the industry, no company offered a special package constructed around an Estonian song festival, although many listed such events in calendars within their literature. The heritage of the 'singing revolution' still appeared to be seen as a matter for expatriate communities, and not for the mainstream Western visitor.

An interest in the Soviet past of Estonia was accommodated by three tour companies. This development appeared to be almost certainly consumer-driven, as the former bases at Paldiski had begun to attract a significant number of independent visitors. This interest in the Soviet past seemed to be common to all three of the Baltic states and has been discussed in the English language press there. One Lithuanian events publication (the *In Your Pocket* guide series) introduced competitions for 'spotting' surviving traces of the occupation. This form of 'heritage' appeared incomprehensible to a large number of Estonians.

The realization that Estonia was part of a far greater tourist region might be considered the most important change in the definition of the tourism product. No fewer than 11 companies now featured a Baltic states' tour. Nine companies included Russia, mainly in the form of packages incorporating St Petersburg, in their literature. These changes appeared to be a response to an obvious consumer demand mediated through the requirements of Western tour operators.

Thus the Estonian tourism product, as perceived by the native tour companies, appeared to be evolving at a rapid rate. It now included a variety of forms of heritage, activities, special attractions, and the addition of the Estonian countryside in its many aspects. What factors could be held responsible for these changes? Some were obviously consumer-led, particularly by the Finnish market. Other influences might be found in foreign participation in the emerging tourism industry, in the form of joint enterprises, subcontracting, and agencies. This participation had important implications for marketing and the definition of the tourist product. It was significant that Estonian Holidays, as a former branch of Intourist, had the most comprehensive product range, perhaps reflecting its past experience and history of dealing with foreign operators. These continuing contacts included Swan Hellenic and Page and Moy, two upmarket British companies specializing in cultural history (M. Saar, 1995, personal communication). The influence of Silja Lines in the Baltic Tours literature was also apparent in the publicity material issued by that company.

Other factors operated at a much lower level. A 'Tallinn Ghost Tour' for example, was devised for Estonian Holidays by an American expatriate. The appearance of Western guide books to the Baltic states might be identified as a possible catalyst for product change and development. It was likely that familiarity with these tourist guides on the part of Estonian tour companies might well have given some indication of Western taste. Guide books to Tallinn, Narva and other towns were available in German as early as 1991, indicative of a curiosity based on historical connections. The *Baltic States* volume of the *Insight Guide* series was typical of those produced in the West, usually American in origin and researched by Americans of Estonian descent with local assistance. The activities of these researchers were quite noticeable in the period after independence and aroused a great deal of interest amongst those in the Estonian tourism industry who were inevitably approached. The new tourist information office in Tallinn received a stream of enquiries by visiting researchers as soon as it opened. The fact that aspects of Estonian heritage were being defined by foreign writers escaped comment at the time.

The demand for guide books aimed at increasing numbers of foreign visitors also had an Estonian dimension. Some native Estonians obviously felt that the holiday product had been restricted and distorted in the Soviet era, and wished to make amends. For example, staff of the Tallinn Pedagogical Institute, already employed by Intourist as summer tour guides, reacted against the previous Soviet establishment-inspired view of Estonian history and culture by producing their own alternative guide book to the capital in 1990 (Anon., 1990). A feeling of local assertiveness and pride animated this sophisticated publication – written in excellent English – which emphasized the problems that a tourist might encounter in what was then still a Soviet society, and apologized for them on the behalf of the Estonian people, often in the form of relaying contemporary national jokes. Unfortunately, this remained the only concrete example of an attempt by Estonian national identity to try to define the tourism product during this period.

It will be seen that the evolution of the Estonian tourism product, in the first instance at least, owed little or nothing to the developing consciousness of a national identity and a national heritage. Instead, it took shape as a result of foreign influences mediated in a variety of ways. Perhaps if independence had come a little later, the vision of an Estonian society created from the bottom upwards might have influenced the perceptions of the personnel of the tourism industry, yet a great gulf of ignorance prevailed on both sides. The members of the Estonian Heritage Society knew little of tourism itself, let alone the theology of 'heritage tourism'. In return, the industry personnel knew little more than their experience of Soviet tourism and its concept of heritage. Any innovation came mainly from the influence of foreign customers and partners, driven by purely commercial considerations. Tragedy played as farce, in which an Estonian tour operator could say with a straight face 'we have folk dancing and silly costumes – that's what brings in the Germans' (Morris, 1992: 5). A profound ignorance of the function of heritage as part of the tourist dynamic on the part of most Estonians seemed to lie at the root of the problem.

The role of Western 'expertise' was sometimes less than helpful – ignorance was to be found on both sides. When the first director of

the Estonian Tourist Board attended an OECD tourism seminar in 1991, she was shocked by the utterances of a Western 'expert': 'Unfortunately, very little is known about the Baltic States…some people thought it would be necessary to develop our "Slavic" culture' (Staff Writer, 1991). This 'expert' had not appreciated that Estonians are not Slavs but Finno-Ugrians. This fundamental misunderstanding symbolized much wasted effort. Expert assistance in the form of 'know-how' funding was usually pitched at the wrong level – dealing with technical training in basic tourism and hospitality skills that the Estonians either possessed themselves or which might be remedied by appropriate courses being instituted in their existing technical and polytechnic institutions. Appropriate theoretical education at a middle and senior management level was neglected.

This is best illustrated in the case of Pärnu College, which was part of the Heritage Society's attempt to create a complete educational system as an alternative to the Soviet model. Indeed, the college soon obtained a national reputation, being the first institution in the country to offer a degree in tourism studies. The curriculum (Finnish in inspiration) was largely a 'business' approach to tourism (Pärnu College, 1995). The missing factor appeared to be tourism theory and analytical techniques that would enable strategic decisions to be made in the long-term. This was reflected in a curriculum which eschewed such concepts as the 'tourism multiplier' and 'sustainability'. Yet the curriculum could usefully be extended to an analysis of 'heritage' as a social dynamic, and contributing to an awareness of the possibilities inherent in coupling the promotion of a living vibrant heritage of the 'singing revolution' to a participatory form of heritage tourism. The result could see synergies that might benefit both Estonian tourism and the future function of heritage as the underpinning of Estonian national identity.

Conclusion

It is thus something of a paradox that Estonians, in the period after regaining independence, having used 'heritage' as a tool of national liberation, should be content to permit the 'commodification' of aspects of their heritage as part of the agendas of foreign tour operators, and to allow aspects of their national identity to be consumer-driven by the demands of international tourism. Yet that, it may be argued, was at the start of the post-Soviet learning curve. What is the current position?

An examination of the current national tourism handbook (Estonian Tourist Board, 2002) reveals only brief mentions of heritage – there is not even a special section on the song festivals. The majority of the websites of the major incoming tour companies hardly feature the topic at all, apart from the usual static heritage attractions and museums. In general, it might appear that little, if anything, has changed. Yet there are exceptions indicating that a reaction to this almost supine attitude to national heritage may be under way.

At the time of independence, there were no restaurants featuring Estonian cuisine in Tallinn, but now there are several. Indeed, beer constitutes something of a national success story. In 1991, tourists were served Swedish and Finnish beers, since it was thought that they would prefer these brands. When, to local surprise, visitors sought out and requested beer brewed in Tartu, Saku and Saarema, entrepreneurial instincts were aroused. The *kaubahall* (supermarkets) began to stock Estonian beers, and visiting Finns soon wanted it made available in Helsinki and elsewhere. The penetration of the Scandinavian market by Estonian breweries is now a source of national pride.

Some Estonian destinations have started to include national heritage in their marketing packages (Worthington, 2003). Pärnu, perhaps the most sophisticated tourist resort in the country, produces a booklet that features a directory of local folk song and dance groups (Pärnu Municipality, nd). Moreover, a new breed of tour operator is emerging, increasingly catering for the 'niche' heritage markets in imaginative interactive ways. The Wris travel company might be cited as an interesting case study (Wris Travel Agency, 2003). At the time of writing it provides a number of specialist packages that are clearly intended to be offered as 'incentive tourism' – as part of a corporate 'bonding activity'. One of these is an invitation to spend a 'soldier's day' in the Red Army in Soviet Estonia, an interesting case of whose heritage ought to be depicted in the new state. However, another option deserves to be quoted in full:

> If you wish to find out what it is like to be an Estonian and to learn about Estonian customs, you should take part in a proper Estonian wedding. We take a look at Estonian life and living, both in the past and the present, build wedding arches made of birch trees, prepare household tasks for the bride and groom to complete, put on some garments characteristic to traditional Estonian folk costumes, and then dance our way to the wedding ceremony. After the ceremony, a proper Estonian wedding feast will be held at a pub, complete with food and drink, singing and dancing, and even more wedding customs, each guest having a special role to play in this joyous festivity! It truly feels good to be Estonian!
>
> (Wris Travel Agency, 2003)

It might be construed as the final irony that Estonian heritage, an important tool of national emancipation, is now pressed into service to reinforce another culture – that of the Estonian and Scandinavian corporate executive.

References

Anon. (1990) *Tallinn – An Alternative Guide*. Revalia Publishing, Tallinn.

Baltic Tours (1992) *Baltic Tours*. Baltic Tours, Tallinn.

Berg, E. (2000) Writing post-Soviet Estonia on to the world map. *Political Geography* 19(5), 601–625.

Estonian Tourist Board (2002) *Estonia Handbook*. Estonian Tourist Board, Tallinn.

Hewison, R. (1987) *The Heritage Industry*. Methuen, London.

Intourist (nd) *The Soviet Baltic Republics*. Intourist, Moscow.

Kihnuranzl Ltd (nd) *Kihnu*. The Open Kihnu Foundation, Parnu.

Lieven, A. (1993) *The Baltic Revolution*. Yale University Press, New Haven, Connecticut.

Morris, P. (1992) Baltics try to boost tourism. *Baltic Independent*, 16 November, p.5.

Palmer, C. (1999) Tourism and the symbols of identity. *Tourism Management* 20(3), 313–321.

Pärnu College (1995) Tourism Degree curriculum. Unpublished MS thesis, Pärnu College, Pärnu.

Pärnu Municipality (nd) *Pärnu Kultuurijuht.* Pärnu Municipality, Pärnu.

Ruutsoo, R. (1993) Transitional society and social movements in Estonia. *Proceedings of the Estonian Academy of Sciences* 42(2), 195–332.

Staff Writer (1991) Travel firms flock to Tourest '91. *Baltic Independent,* 13 November, p.3.

Taagepera, R. (1993) *Estonia: Return to Independence.* Westview Press, Boulder, Colorado.

Urry, J. (2001) *The Tourist Gaze,* 2nd edn. Sage, London.

Worthington, B. (2003) Change in an Estonian resort: contrasting development contexts. *Annals of Tourism Research* 30(2), 369–385.

Wris Travel Agency (2003) *Activities – incentive.* Wris Travel Agency, Tallinn <http://www.wris.ee/eng/activities/index.html>.

7 Sustainability as the basis for future tourism development in Serbia

Jovan Popesku and Derek Hall

This chapter examines Serbia's pursuit of a 'sustainable' tourism development policy and the context within which this is evolving. This context comprises:

- Serbia being set in relative isolation during the 1990s;
- the physical destruction and regional instability wrought by the break up of Yugoslavia; and
- the consequent loss of markets and the need to reassert a new identity both for Serbia and its regional setting.

Out of the ruins of empires destroyed by World War I, the Kingdom of Serbs, Croats and Slovenes, later to be Yugoslavia, was created as an uneasy mix of rival south Slavs as well as non-Slav groups. Spanning a cultural fault line and acting as a buffer state between East and West in Europe, the country's political interests were strongly influenced during the inter-war period by Yugoslavia's largest constituent groups – Serbs and Croats.

Out of World War II, Communist partisans under Josip Broz Tito re-established a Yugoslav state in 1944, but by trying to steer a relatively independent path within an ever-enlarging Soviet-dominated bloc, Tito caused Stalin to expel Yugoslavia from the Cominform organization in 1948, thus effectively isolating Yugoslavia from its fellow Communist states in Europe.

Thereafter, Yugoslavia was required to steer a delicate domestic and international political and economic path. By the mid-1960s, a new constitution was devolving wider powers to six states and two autonomous regions within the federal republic. Yugoslav citizens were permitted to live and work abroad, and a much greater degree of private and foreign investment in such economic sectors as tourism was possible.

The Adriatic Highway, running virtually the whole length of the Yugoslav coast, was completed between 1961 and 1965, with Western assistance. Although primarily intended to integrate often isolated communities, the completion of this access link stimulated coastal tourism development and projected Yugoslavia's littoral to the forefront of international interest and investment in the country (Allcock, 1991; Bakić and Popesku, 1992). By the mid-1970s emigré workers' remittances and tourism receipts represented by far the two most important sources of foreign currency for Yugoslavia as a whole, even though these trends largely impacted on the north-westerly republics of Slovenia and Croatia.

Land-locked constituent republics such as Serbia received relatively little direct benefit or tourism experience, although a redistribution of state income attempted to mitigate economic imbalances, but rural depopulation and migration to the coast only acted to allay rural development in a number of interior regions.

The increasingly complex series of two-way linkages with Western Europe, and notably with German-speaking countries, provided the context within which Yugoslavia proceeded to cement a national identity in the West around experiences of inexpensive coastal mass tourism, reinforced with the images of such cultural-historic icons as the medieval walled city of Dubrovnik. However, such identity was derived largely from the attributes of the narrow Adriatic strip of the country steeped, untypically for Yugoslavia as a whole, in Roman and Venetian heritage. The country's often rugged interior – including its major cities – was largely excluded from the tourist mental map. Exceptions to this were the lakes and mountains of Slovenia, immediately accessible to the Austrian, north-western Italian and southern German markets, and winter sports centres stimulated by the hosting of the winter Olympic Games in Sarajevo (Bosnia-Hercegovina) in 1984. Apart from the urban attractions of Belgrade, again Serbia was largely excluded from this tourism map. Yet, 'Yugoslavia' was so successful in its attraction of mass tourism markets that by the late 1980s the country was generating more tourism hard currency income than the rest of communist eastern Europe combined (Hall, 1998).

During the first half of the 1990s, while conflict traumatized various parts of the former Yugoslavia, and economic development and especially tourism was put on hold, post-communist transformation in much of the rest of Central and Eastern Europe proceeded apace. This was just at a time when smaller-scale specialized niche tourism was receiving global stimulation, providing opportunities for small indigenous companies to develop high value-added products. Through the 1990s, this dimension intensified as tourism, alongside other service and manufacturing industries in Central Europe, gained experience of responding to changing 'Western' market demands.

Tourism as Re-imaging National Identity

In the former Yugoslavia as a whole, part of the escape strategy from the past and its images has been to diversify away from mass, largely coastal, tourism and to emphasize the uniqueness of cultural and natural resources. This has provided rejuvenated opportunities for land-locked territories such as Serbia. It has

entailed an implicit shift in geographical emphasis to promote countries' interiors and in a structural refocusing to emphasize products rather than destinations. Rural and nature tourism products provide an important element of this reinvigoration, with professional marketing undertaken by local and central government, NGOs and private sectors with the assistance of Western consultants. Tourism promotion literature has become notably niche-oriented (e.g. NTOS, 2000; Slovenia Tourist Board, 2000; Stifanic, 2000). This has entailed the re-imaging of many rural areas and a greater product focus within frameworks portraying sustainability (Roberts and Hall, 2001).

'Heritage' occupies a more ambivalent role in a region with a diversity of cultures and histories, and where the past is often drawn on to justify the present. 'Heritage' is far from being a value-free concept: economic power and politics influence what is preserved and how it is interpreted (Lowenthal, 1997). Promotion of secular rural and urban heritage, as an integral element of cultural history, was characteristic of the communist period. Under the communists, however, such heritage was employed to overcome or subdue ethnic rivalries and as a means to inculcate a unified sense of Yugoslav identity and pride.

Nature tourism is appealing in that not only is it a potentially high-spending sector of relatively passive, low-impact tourism, which can encompass wildlife viewing and 'ecotourism', but its very promotion and pursuit can convey for the image of the host destination notions of conservation, an unpolluted healthy environment and life-style, sustainability, ethical sensitivity and a relative cultural and political (value-free) neutrality, important in a region with much cultural and political baggage.

Attractive landscape or particular elements of the natural environment can provide an important attraction in conjunction with other activities. In forest and wilderness areas the attraction of isolation and solitude paradoxically appeals to many visitors. Water bodies can play an important role in enhancing a landscape. Lakes and rivers both add visual beauty and offer the possibilities of swimming, sailing, canoeing and fishing, providing alternative locations to sea coasts. The popularity of Lake Ohrid, shared between Albania and FYR Macedonia, and Lake Bled in Slovenia is notable in this respect. The spectacular scenery of the Carpathian mountains in Romania, or the Rila/Pirin ranges in Bulgaria, enhances winter sports resorts and draws visitors during the summer months for walking and climbing.

Nature tourism has received substantial support in Central and Eastern Europe since the mid-1990s, with professional marketing undertaken by local and central government, NGOs and private sectors (e.g. Bognar, 1996; Witak and Lewandowska, 1996). Promotion is, however, very variable. For example, low-key marketing characterizes Sakertour, the ecotourist branch of the Greenscape Company, comprised of a group of young English-speaking Hungarian and Slovakian ornithologists, which has been organizing tailor-made professional ornithological trips and field studies in the Carpathian Basin since 1994 (Sakertour Greenscape, 1998; Sakertour, 2002).

However, at a destination level, any attempt at image-building faces at least three major constraints:

- a lack of adequate finance to support appropriate marketing campaigns – a common trait – has been exacerbated by limited experience of global markets and a lack of appropriate expertise: pre-1989 mass markets were drawn from a relatively small number of West European countries;
- second, in a post-conflict context where rapid reconstruction and re-imaging is sought, tourism destination marketers may be pressured to return short-term results when long-term investment is required to build a consistent brand. Such a dilemma may be reflected in message inconsistency;
- third, while centralized, relatively authoritarian regimes could impose some control and coherence over the component parts of a destination image, it is more difficult within post-communist market economies to develop a coherent image for destinations which are composed of a large number of products and environments.

Recent Trends in Tourism Development in Serbia

The decrease of the tourist flow, particularly foreign, as well as the decrease of foreign currency income generated from tourism, is the obvious indicator of the causes and the consequences regarding the condition and position of the Serbian tourism industry, especially the hotel industry.

Largely as a result of national and regional political change and conflict, between 1989 and 2000, the number of tourist overnights in Serbia decreased by 36.6%, with a 71.6% decline in international tourist overnights. The decrease of total tourist arrivals in the same period was 47.9%, of which arrivals of domestic tourists decreased 37.8% and of foreign tourists 82.4%. The foreign currency income generated from tourism was reduced by 93.6% in the period 1990–2000. In order to express the scale of lagging behind this world tourism trend, it should be stressed that in the period 1989–2000 the total tourist flow in the world, expressed by the number of arrivals, increased by 73.1%, and the total income generated from international tourism increased by 127.7% (Unković, 2001; Popesku, 2002).

Tourism's share of gross domestic product remained at 2%, while its share of the number of employees actually increased from 3 to 4%, largely because of structural changes in the Serbian economy over the period. It is especially important to emphasize the fall of the share of the tourism industry in the total investments from 2.3% to 0.5%, as well as in the foreign currency income from 5.8% to 1.7%. The average use of the total accommodation capacities was 24.8% (Marić, 2001).

Serbia's role as an international tourism destination was rejuvenated by FR Yugoslavia's (Serbia and Montenegro) re-admission to international institutions and normalization of international relations, following democratic change – a 'liberal revolution' (Podunavac, 2002) – within the country. During 2000 and 2001, an increase in international tourist arrival numbers was notable, especially to the larger cities and particularly to Belgrade.

For a decade, because of internal and external political and conflictual circumstances, tourism for Serbia represented a missed opportunity. Indeed one may recognize surprising parallels with other countries examined in this volume, such as South Africa (Chapter 14). Ironically, during this period the tourism industry was significantly protected – albeit in a negative sense – from foreign market competition, while conversely, foreign investment in Serbian tourism facilities was severely limited. Positively, tourism resources were largely protected from the degradation and erosion of international mass tourism demands. However, the environmental impact of NATO bombing during the Kosovo conflict was significant (Ashford and Gottstein, 2000; Ott, 2000).

Serbia: a Brief SWOT Analysis

The present balance of positive and negative factors influencing the development of tourism in Serbia can be summarized within a conventional SWOT (strengths, weaknesses, opportunities, threats) analysis (Popesku, 2002).

Strengths

- Diversity of tourism products (mountains, e.g. the winter sports centre of Mt Kopaonik, spas, lakes and rivers, city centres, notably Belgrade, rural tourism, potentials for ecotourism, adventure tourism);
- well-preserved natural environment;
- richness of cultural heritage (medieval monasteries: Studenica and Sopocani monastery are on UNESCO list of world cultural heritage); and
- hospitality of the people.

Weaknesses

- Long-term absence of Serbia from the international tourism market due to the effects of the economic sanctions and the isolation measures enforced by the international community, as well as self-isolation, resulting in a poor national image in Western markets;
- insufficient market recognition of Serbia as a tourism destination – low-profile image and/or negative image caused by the non-tourism factors:
- long-term underestimation of the realistic development opportunities by the official institutions responsible for the formulation of tourism development conditions, and therefore, as a result, insufficient investment in the development and the promotion of this field of the economy;
- insufficient level of privatization in the tourism industry, especially in the hotel industry;
- inappropriate involvement and disproportionate influence of some local communities and certain interest groups in tourism development;
- lack of both general and specialized tourism infrastructure, especially in the less developed areas of the country;
- lagging behind world trends in the sphere of education for tourism professionals (international accreditations, licences, etc.).

Opportunities

- The range of tourism products already known to the international tourism market (Belgrade, Kopaonik, monasteries) and an ability to identify and target market segments for these products;
- favourable/changed conditions in the surrounding region, favouring Serbia's return to the tourism market;
- beginning of the process of the privatization and presence of international hotel chains in Serbia (Hyatt, Intercontinental, Best Western); and
- generally, national and international tourism and associated factors have become more favourable for the return of Serbia to the tourism market.

Threats

- Absence of appropriate cooperation with foreign tour operators;
- lack of appropriate national, regional and local structures for the development, management and promotion of tourism;
- long-term decline of tourism services' quality; and
- inappropriate managing of the environment in relation to tourism.

Serbia is therefore at a turning point in establishing a new beginning: re-defining itself as an international tourist destination. In recent years there has not been a clear, or at least positive, image of Serbia in this respect. However, in February 2003 the name 'Yugoslavia' was formally abolished, with the country's constituent parts – Serbia and Montenegro – agreeing to remain in a loose federal relationship for 3 years. This at least will remove one element of international confusion over Serbia's identity. Three interrelated priorities must therefore be:

- to establish Serbian tourism in a strong position within the international market: to this end, the need for tourist visas was abolished in June 2003 for visitors from Western and Central Europe, Israel, North America, Australia, New Zealand, Singapore and the Republic of Korea, as well as Slovenia and Croatia (NTOS, 2003);
- to target new markets and promote a new positive brand image for Serbia; and
- to emphasize sustainable development as an ethos underlining all elements of coordinated and balanced tourism development in Serbia, enhancing the richness of the country's natural and cultural heritage.

Defining the Strategic Frame for Sustainable Tourism Development in Serbia

The new domestic democratic context, coupled with a striving to attain legal and other forms of adaptation to European Union norms, represents an important starting point for infusing sustainability principles into the processes of tourism development in Serbia. To establish a strategic framework for this, certain prerequisites are needed:

- a comprehensive registration of all tourism resources;
- substantial, appropriate and well-resourced market research; and
- a critical analysis of existing laws and other regulations in the field of tourism to evaluate the requirements for harmonization with international, and particularly European Union, standards.

A number of positive preconditions already exist:

- the law on tourism (1994);
- revisions to the law on tourism (1999);
- various acts based on the law on tourism (1994–1997);
- the Spatial Plan of the Republic of Serbia (1995); and
- a national strategy for tourism development to 2010, which was adopted in 1999.

The government has also prepared a new law on tourism (2003) in order to harmonize with European Union legislation and to be able to adapt to global conditions in tourism development.

The key principles for the implementation of a sustainable development policy in tourism management were emphasized in the 'tourism development strategy to 2010' document (Čerović, 2002). This clearly states that co-ordinated and sustainable development, aligned to a concept of limited growth, represents the only realistic approach for tourism development in Serbia. This is especially important given Serbia's range of important non-renewable natural resources which can act as the basis for future tourism – and particularly ecotourism – development (Popesku and Marić, 2003). It will be crucial for the sustainability of tourism to achieve coordination of the development process and to preserve the quality of the natural environment (Government of the Republic of Serbia, 1999).

Sustainability can only become the basis of appropriate tourism development in Serbia if its principles are included in the process of strategic planning and can be expressed in strategic goals. Defining the strategic framework for tourism development requires the establishment of appropriate procedures and organization. It would seem necessary for this purpose for the government, through the ministry responsible for tourism, to establish a tourism task team consisting of the representatives of government, ministry responsible for tourism, national tourism organization, local authorities, tourism industry, trade unions and non-governmental organizations. The objective of this body should be to formulate the draft of the basic strategy for sustainable tourism development with particular emphasis on the development of ecotourism. This process should be based on scientific methods of research and analysis, consultation with appropriate experts, critical evaluation of the results, and should encompass the convening of national tourism workshops in order to share and stimulate debate.

The definitive strategic framework should be in balance with the development of the infrastructure capacity as well as with the development of complementary economic and non-economic activities. The strategic framework should provide the context for a national tourism development plan, incorporating a long-term master plan as well as medium-term programmes for national tourism development and for specific destination regions. The basic aim of this approach should be the striving for a balance between economic, socio-cultural and environmental goals, which require it to be based on sustainable development principles. The leading role of government in creating and defining the conditions for tourism development should be complemented by partnership between the main stakeholders: government at all levels, the industry and its associations, trade unions, local populations and non-governmental organizations.

High quality, well-preserved nature is one of the fundamental competitive advantages for positioning Serbian tourism on the international tourist market (e.g. see Cok, 1998). The variety of natural resources available points to the potential of ecotourism as an important tool for sustainable tourism development. An important basic resource for the development of ecotourism as a niche market is Serbia's protected areas. Currently 5% of Serbia's territory is regulated for nature protection, but plans aim to extend this to 10% by the year 2010.

Such protected areas encompass:

- five national parks: Djerdap (the Iron Gates on the Danube in eastern Serbia), Mt Fruska Gora (in Vojvodina province), Mt Kopaonik (2017 m asl), Šar Planina and Mt Tara (1544 m asl, in western Serbia) (NTOS, 2001a);
- three Ramsar sites: Lake Ludaš, 593 ha, designated in 1977; Obedska Bara, an oxbow lake and surrounding area, 17,501 ha, 1977; and Carska Bara – ponds, river, swamp, marsh, 1767 ha, 1996 (Ramsar, 2003);
- one biosphere reserve – 'Golija-Studenica' – officially approved by the MAB Board in 2001 (UNESCO, 2003); and

- 120 nature reserves, 20 nature parks and about 470 natural monuments under protection, including nine major cave systems, as well as 215 plant and 427 animal species designated as natural rarities (Institute for Protection of Nature of Serbia, 2003), with 140 bird species of European conservation concern (Fat Birder, 2003).

To complement an effective policy of developing sustainable forms of tourism in Serbia, it is necessary to recognize both domestic and global constraints which need to be overcome. It is important to understand the processes of globalization, particularly in relation to accelerated integration of economic activities in much of the rest of Europe, a process which is rapidly changing the nature of relationships in international tourism and other cross-border activities (Saayman, 2001).

The strategic goals of national tourism development demand that it is necessary to change the image of Serbia as an international destination. The prevailing image can be largely characterized as at best neutral, and mostly negative. This reflects a decade of hostile international media attention and regional conflict and of course is not the result of tourism factors.

Tourism Image Requirements

Serbia's requirement for brand development is evident at a number of levels. The identity-building requirements demanded by the global marketplace are essentially:

- psychological: to help establish a positive and appropriate image which will appeal to tourists, promoters and investors;
- economic: the need to respond to new and changing market demands and increasing market differentiation – by generating customer loyalty through repeat visits, and raising per capita tourism income levels through targeted quality product development; and
- political: the need to disassociate from the recent past and from regional instability, and to emphasize a 'Europeanness'.

European economic and political convergence, and the leisure search for new experiences and products, provide a potentially wide range of opportunities for the interweaving of national branding and the promotion of tourism. Although better knowledge of markets is required there is also the need to respond to the new opportunities opened by technological development. Significantly, much effort is being put into producing lively and imaginative websites, with marketing messages and logos which are emerging as branding tools in their own right. However, there may be historical dimensions which will continue to constrain Serbia's ability to take advantage of the potentially wide range of markets opened to them. This is partly due to regional instability and to characteristics which have appealed to a relatively stable and limited range of markets.

National Identity and Political Change: Promoting Serbia

While images of heritage have been employed as a vehicle for restructuring and modernization particularly in rural regions (Dewailly, 1998; Light, 2000), in Serbia, since the mid-1990s, the combination of nature, identity and tourism has been expressed in the promotion *Serbia: Landscape Painted from the Heart* (e.g. NTOS, 1996), a strapline derived from a well-known Serbian quotation. In terms of heritage, however, the cultural landscape portrayed revealed an exclusive concentration on Serbian/Orthodox tradition (Hall, 2002).

The challenge for Serbia's image-promoters and tourism marketers is to balance, on the one hand, positioning the country in a favourable context in relation to major ('Western') tourism markets and sources of inward investment, and on the other hand exploiting the heritage of a majority Christian Orthodox culture which has appealed to Serbia's hitherto political and tourism supports – Greece, Bulgaria and Russia.

Indicative of change, one of the first actions of National Tourism Organization of Serbia in 2001 was to adopt a new logo and strapline as a tool in re-positioning Serbia as an international tourist destination. Saatchi & Saatchi Advertising Balkans (also the 'creators' of *Landscape Painted from the Heart*) produced the message *Fall in Love Again – Serbia: Three Times Love*.

Fall in Love Again is primarily directed at those who had previously experienced Serbia as a tourism destination, particularly before 1989. These fell into three categories:

- citizens of the former SFR Yugoslavia (Slovenia, Croatia, Bosnia-Hercegovina, Macedonia);
- residents of neighbouring countries (Romania, Bulgaria); and
- tourists from countries that had been relatively successful markets for Serbia in the 1960s–1980s (Germany, Italy).

As one of the strategic goals for 'tourism development to 2010' is an orientation to neighbouring countries' regional and European markets, this slogan can be understood as an open-hearted invitation to experience again holidays in Serbia. On the other hand, it is also an invitation addressed to all potential visitors to come to Serbia and to feel emotions that may have been lost in normal everyday life.

For the National Tourism Organization of Serbia *Three Times Love* represents nature, culture and people: three crucial factors that are fundamental for tourism development in Serbia – well-preserved natural environment, richness of cultural heritage and the hospitality of people. This is also an invitation for tourists:

- to enjoy themselves amongst beautiful preserved nature, national parks and natural reserves and well-protected flora and fauna;
- to visit the great variety of cultural and historical monuments; and
- to meet Serbian people, to feel their warmth and hospitality, to discover their customs, folk art and way of life and to enjoy delicious healthy cuisine (NTOS, 2002).

Accompanying the *Serbia – Three Times Love* strapline in the new Serbian brand promotion is a symbol in the form of a stylized representation of a window from the country's much-visited 12th-century Studenica monastery (NTOS, 2001b), which is comprised of three parts. Three also suggests the Holy Trinity and reminds us that 80% of the citizens of Serbia are Christians (e.g. see Djurdjev, 2000).

There is no other special significance in the *Three Times Love* slogan: for example it holds no references to Serbian literature, unlike the former slogan *Landscape Painted from the Heart* (1995–2001) which was actually a paraphrase of the title of a well-known novel (*Landscape Painted by Tea*) by the well-known Serbian author Milorad Pavic.

Unfortunately the slogan fails to indicate directly any sustainability emphasis, although it implicitly underscores the preservation of natural and cultural heritage and involvement of the community in the creation of tourism products. But for this shortcoming alone the slogan should be revised. Further, its concept is not clear for consumers, and as explanation of its meaning and significance is needed, the main objective of the slogan is not realized. It is to be hoped that a clear message of sustainability and inclusion will be developed as a foundation for future tourism development in Serbia and for the prosperity of a multi-ethnic Serbian state.

Conclusions

Re-integration into the global economy and preparation for eventual accession to the European Union are now prime foreign policy driving forces for Serbia. International tourism has been recruited to play a significant role in this process – both in terms of image reinforcement and as a vehicle for international 'normalization', collaboration and benchmarking. However, Serbia's positioning strategy still has much potential for development, having been constrained by a combination of lack of finance, experience and expertise coupled to lingering regional (and global) instability.

In tourism terms, a change in emphasis from a national destination to a product focus will enhance appeal in the pursuit of sustainable niche segmentation, but this may be a difficult task to undertake in the short to medium term. Further, consumer research is needed to identify what are potentially the most motivating propositions for tourism markets. The value of an established brand lies not least in the perceptions of consistency and quality that it represents. Collaboration, networks and partnerships are essential in assisting the generation of appropriate products and product images if Serbia is to position itself in relation to key international markets and investors.

References

Allcock, J.B. (1991) Yugoslavia. In: Hall, D. (ed.) *Tourism and Economic Development in Eastern Europe and the Soviet Union.* Belhaven, London, pp. 236–258.

Ashford, M.W. and Gottstein, U. (2000) The impact on civilians of the bombing of Kosovo and Serbia. *Medicine, Conflict and Survival* 16(3), 267–280.

Bakić, O. and Popesku, J. (1992) Marketing of Serbia as a Tourist Destination. International Conference *Tourism in Europe – The 1992 Conference*, Newcastle-upon-Tyne.

Bognar, A. (1996) *Budapest's Protected Natural Heritage.* Municipality of the City of Budapest, Budapest.

Čerović, S. (2002) *Strategijski Menadžment Turističke Privrede Srbije.* Želnid, Belgrade.

Cok, V. (1998) Legal grounds of environmental protection in Yugoslavia. *Yugoslav Law* 25(3), 159–174.

Dewailly, J.-M. (1998) Images of heritage in rural regions. In: Butler, R., Hall, C.M. and Jenkins, J. (eds) *Tourism and Recreation in Rural Areas.* John Wiley & Sons, Chichester, UK, pp. 123–137.

Djurdjev, B.S. (2000) Problems of regional population growth in Serbia. *GeoJournal* 50(2-3), 133–138.

Fat Birder (2003) *Birding Serbia and Montenegro.* Fat Birder <http://www.fatbirder.com/links-geo/europe/serbia.html>.

Government of the Republic of Serbia (1999) *The Strategy of Long-Term Tourism Development in Serbia until 2010.* Government of the Republic of Serbia, Belgrade.

Hall, D. (1998) Tourism development and sustainability issues in Central and South-eastern Europe. *Tourism Management* 19(5), 423–431.

Hall, D. (2002) Branding and national identity: the case of Central and Eastern Europe. In: Morgan, N., Pritchard, A. and Pride, R. (eds) *Destination Branding – Creating the Unique Destination Proposition.* Butterworth-Heinemann, Oxford, UK, pp. 87–105.

Institute for Protection of Nature of Serbia (2003) *Nature Reserves.* Institute for Protection of Nature of Serbia, Belgrade <http://www.natureprotection.org.yu/english/reserves.php>.

Light, D. (2000) Gazing on communism: heritage tourism and post-communist identities in Germany, Hungary and Romania. *Tourism Geographies* 2(2), 157–176.

Lowenthal, D. (1997) *The Heritage Crusade*. Viking, London.

Marić, R. (2001) *Strategija Razvoja Turističke Industrije Srbije do 2010. Godine*. Ekonomski Institut, Belgrade.

NTOS (National Tourism Organization of Serbia) (1996) *Serbia: Landscape Painted from the Heart*. National Tourism Organization of Serbia, Belgrade.

NTOS (National Tourism Organization of Serbia) (2000) *Landscape Painted from the Heart*. NTOS, Belgrade <http://www.serbia-info.com/ntos/inf_gen.htm>.

NTOS (National Tourism Organization of Serbia) (2001a) *Natural Attractions*. NTOS, Belgrade <http://www.serbia-tourism.org/engeski/nature/atr_menu.htm>.

NTOS (National Tourism Organization of Serbia) (2001b) *Welcome: Republic of Serbia*. NTOS, Belgrade <http://www.serbia-tourism.org/>.

NTOS (National Tourism Organization of Serbia) (2002) *Press Releases – International Tourism Fairs*. NTOS, Belgrade.

NTOS (National Tourism Organization of Serbia) (2003) *News*. NTOS, Belgrade <http://www.serbia-tourism.org/engleski/vesti/evesti.htm>.

Ott, E. (2000) Kreig und Umwelt in Jugoslawien. *Prokla* 30(3) 469–482.

Podunavac, M. (2002) Revolucija i ustav u postkomunizmu: slucaj Srbije. *Politicka Misao* 39(1), 128–138.

Popesku, J. (2002) Ecotourism – A Tool for Sustainable Tourism Development in Serbia. Paper presented at *The World Economic Summit*, Quebec City.

Popesku, J. and Marić, R. (2003) Ecotourism Development as a Tool for Repositioning of Serbia as a Tourist Destination. Paper presented at the International Scientific Conference *The Development and Potentials of Ecotourism on Balkan Peninsula*, Pirot, Serbia, 30 January–2 February.

Ramsar (2003) *Ramsar Sites Database*. Ramsar, Gland <http://www.wetlands.org.RDB/Ramsar_Dir/Yugoslavia_.htm>.

Ritchie, J.R.B. and Crouch, G.I. (2000) The competitive destination: a sustainable perspective. *Tourism Management* 21(1), 1–7.

Roberts, L. and Hall, D. (2001) *Rural Tourism and Recreation: Principles to Practice*. CAB International, Wallingford, UK.

Rosić, I. and Popesku, J. (eds) (1999) *Ključni Aspekti Razvoja Seoskog Turizma Srbije*. Ekonomski fakultet, Kragujevac.

Saayman, M. (2001) Tourism growth and tourism policy: the case of an emerging market. In: Keller, P. and Bieger, T. (eds) *Tourism Growth and Global Competition*. AIEST, St Gallen, pp. 131–148.

Sakertour (2002) *Sakertour: the Carpathian birding company*. Sakertour, Debrecen <http://www.sakertour.hu>.

Sakertour Greenscape (1998) *Birdwatch Hungary 1998*. Sakertour Greenscape, Debrecen.

Slovenia Tourist Board (2000) *Slovenia by Bicycle*. Slovenia Tourist Board, Ljubljana.

Stifanic, D. (2000) *Vrsar: Bike Ecoride*. Tourist Association of Vrsar, Vrsar.

Unković, S. (2001) *Ekonomika Turizma*. Savremena administracija, Belgrade.

UNESCO (2003) *UNESCO – MAB Biosphere Reserves Directory*. UNESCO, Paris <http://www2.unesco.org/mab/br/brdir/directory/biores. asp?mode=all&code=YUG+02>.

Witak, A. and Lewandowska, U. (1996) *Poland: the Natural Choice*. Sport i Turystyka, Warsaw.

8 Positioning the tourism product of an emerging industry: image, resources and politics in Kyrgyzstan

Peter Schofield

Introduction

The Kyrgyz Republic's landscape is dominated by its mountains that cover 94% of its 198,500 sq km land area (Fig. 8.1). Not surprisingly, this natural topography features prominently in both the country's tourism product and its 'official' promotional literature, created and projected by the Kyrgyz State Agency of Tourism and Sport (KSATS). The physical terrain has also had a significant influence on the country's political, economic and social development in that the mountains have isolated the state and hindered the development of internal transportation routes. The Fergana range divides the country into discrete northern and southern regions with distinct clan, ethnic, religious and political affiliations; this has constrained government attempts to forge a national identity for Kyrgyzstan and a coherent image and brand for its tourism product – key factors affecting the establishment of a clear and distinctive position for the Republic.

Since gaining independence from the Union of Soviet Socialist Republics (USSR) in 1991, after 72 years of Soviet rule, the Kyrgyz Republic has tried to develop a post-colonial identity from its Soviet, Turkic and Kyrgyz heritage (Figs 8.2–8.5) while attempting to differentiate itself from its neighbours, China, Kazakhstan and Uzbekistan, politically through democratization and on an economic basis by

introducing market principles in an attempt to become a 'Switzerland in Central Asia' (Fletcher and Sergeyev, 2002). Tourism has become a focus for state policy because of the opportunities it provides for the pursuit of economic goals (Williams and Shaw, 1998). In addition, tourism's political potential has been recognized with respect to facilitating the development and projection of a national identity and creating a favourable position for the Republic in the international marketplace. Currently, the country's tourism product is 'underpositioned' (Baker, 1992) because awareness of the destination in the world's major tourist-generating areas is poor and/or indistinguishable from other Central Asian countries. The creation and transmission of an appropriate image of Kyrgyzstan is therefore an important issue, given that a destination's image is a key component of the product (Coshall, 2000; Tapachai and Waryszak, 2000) and that consumer perceptions of tourist destinations are profoundly influenced by the 'official images' of place (Gunn, 1974) projected by National Tourism Organizations (NTOs), particularly where exposure to mass media images of such destinations has been limited (Court and Lupton, 1997).

At this embryonic stage in Kyrgyzstan's place-branding process, it is important to identify the key issues that are affecting the development of both the country's image and that of its

Fig. 8.1. Kyrgyzstan.

Fig. 8.2. Two faces of Kyrgyzstan: Turkic and Soviet heritage in Bishkek.

Fig. 8.3. Kyrgyzstan: settlements on the outskirts of Bishkek.

Fig. 8.4. Kyrgyzstan: traditional transport: the horse culture of Central Asia.

Fig. 8.5. A traditional Kyrgyz yurt at Lake Song-Kyol.

tourism industry to provide an analytical framework. This chapter therefore examines the effect of political, economic, social and cultural factors on the emerging tourism product and its positioning. Within this context, there are two central concerns. Firstly, the nature of the relationship between government policy, the construction of national identity and the development of tourism. Secondly, the effects which Kyrgyzstan's rich cultural heritage and ethnic diversity are having on the development of an appropriate image for the Republic. In tourism development terms, this is both a blessing and a blight. On the one hand, 'cultural otherness' (Wood, 1997) – its perceived cultural 'differentness' (Hitchcock *et al.*, 1993) – is an important aspect of place brand equity that augments the country's natural features and has the potential to evoke an emotional attachment. On the other hand, the awakening of clan, ethnic and religious identity, particularly marked in the south of the country, militate against the development of a national identity and a coherent tourism product and image.

One of the immediate problems of researching tourism in Kyrgyzstan is the absence of up-to-date, reliable and consistent statistics and the lack of previous research on tourism issues. This is common in former Soviet states (Hall, 2002). The analysis has therefore been based

on international economic data, government agency statistics, press, trade and journal articles and the country's 'official' promotional literature. However, many of the facts were difficult to corroborate and figures tended to vary by up to 30% depending on the source. Even estimates of Kyrgyzstan's population ranged from 3.5 to 5 million, although the official figure issued by the National Statistical Committee of the Kyrgyz Republic (NSCKR), based on the Republic's first census conducted in 1999, is 4.85 million (NSCKR, 2002).

Post-Soviet Politics

The disintegration of the Soviet Union, the emergence of five independent Central Asian republics (Kazakhstan, Tajikistan, Turkmenistan, Uzbekistan and Kyrgyzstan) and the formation of the Commonwealth of Independent States (CIS) brought the region to the world's attention because of its strategic importance from a geo-political perspective. Central Asia's subsequent political and economic evolution and rapidly growing population, together with events in the aftermath of 11 September 2001, have focused that attention because these countries are returning to the languages, culture and traditions of their pre-Soviet past, including a resurgence of both their common religion, Islam, and clan-style politics. After gaining its independence, Kyrgyzstan faced similar socio-economic and political problems to other CIS states, typical of low-income developing countries. What differentiated Kyrgyzstan from its neighbours was its distinct development path – a pro-Western re-orientation to establish a free market economy and a democratic society (Starr, 1996).

There is a general consensus in the political development literature that something more than economics is necessary for democracy to be established and maintained (Lipset, 1994; Miller *et al.*, 1994; Finifter, 1996; Linz and Stepan, 1996; Huskey, 1997; Gibson, 1998; Anderson, 1999). Theoretically, within the context of post-communist transition, a link between economic development and democratization might be expected. The introduction of the private sector and the diversification of ownership in both industrial and agricultural sectors would theoretically create new interests that gradually develop their own organizational and social networks, creating cross-cutting identities that would mitigate the worst effects of more exclusive ethnic, tribal or religious ones (Anderson, 2000). In practice the situation is more complex, and in Kyrgyzstan social organizations have struggled to develop or compete with more traditional power networks. The Democratic Party, the Republican Party and the National Rebirth Party have attempted to fill the vacuum left by the demise of the Communist Party, but it is the two pre-eminent clans, Salto and Saribagish, that dominate state politics. Whilst there is a tolerance of diversity that stems from the country's hospitable tradition, there is considerable suspicion about many of the consequences of the new pluralism, particularly in the religious sphere. Many interest groups, particularly in the south, have not been supportive of either democratization or Akayev's policy of 'selling out to the West'. This reflects the close links between Islam and the state (Lipset, 1994) and widespread misgivings about attempts to impose Western notions of a civil society, rooted in individual choice, on a country where group identities are more potent and there is a natural inclination to look for strong leadership within the community and state.

It is therefore not surprising that in 1994, President Akayev appeared to step back from his earlier belief in the simultaneous development of political, economic and social reform when he stressed that democratization would be an evolutionary process during which it would be necessary to strengthen the capacity of the central state and grant the executive more power to push through reform. Opinion is divided on the subject of Akayev's more authoritarian policies. One school of thought believes that the step to the right is a temporary measure and a political necessity to overcome opposition to economic reform and ensure a measure of social and political stability during this turbulent period in the country's evolution. Others perceive that Akayev has become disillusioned with the classical democratic path to development because Kyrgyz history, geography and politics have resulted in sub-optimal outcomes from seemingly laudable policies. He has therefore adopted solutions more suited to Kyrgyzstan's idiosyncrasies (Huskey, 1995; Nellis, 1999). Clearly, the development of a modern civil society in Kyrgyzstan is a key factor on its road to democracy. What is less clear is whether domestic interests or Central Asian neighbours will permit the Republic to consolidate democratization, without which it will struggle to attract international attention and achieve its integration into world markets.

The country is experiencing difficulties arising from the complex change in ideology which privatization and competition have brought (Anderson, 1999). The government's attempt to transform its planned economy into a free market in which the private sector plays a decisive role has proved to be a particularly difficult process because of the severe economic and social problems it faces. Not surprisingly, the country's politics bear the imprint of economic backwardness and dependency; in addition, they reflect the Republic's acute ethnic and religious cleavages and its inter-clan rivalries that have been smouldering for centuries (Huskey, 1995).

The Economic Context

The five newly independent Central Asian states have a geographic area of approximately

670,000 square miles (40% of the size of the USA), extensive mineral reserves and a population of over 50 million, which represents a substantial market. From an economic perspective, the five countries are heterogeneous. Despite having a primarily agricultural base, they vary significantly with regard to natural resources. Kazakhstan has vast oil reserves, including the Tengiz field, which is the world's largest known unexploited oilfield, Turkmenistan is rich in natural gas, Tajikistan has the fourth largest aluminium production plant in the world and Uzbekistan is one of the world's major cotton producers. By comparison, Kyrgyzstan has an abundance of rare minerals such as gold, caesium, mercury and antimony, but is one of the poorest Central Asian states. This results from a number of factors including its largely unproductive high ground, its underdeveloped hydroelectric power capability and its seriously unbalanced and overspecialized industrial base, the legacy of serving the Soviet military–industrial complex – one of the least efficient sectors (Bruder and Scanlan, 1993). Kyrgyzstan currently trades with over 50 countries, although most trade is with former Soviet countries. Kazakhstan and Russia supply 50% and 30% of the Republic's petroleum, respectively (Banks and Ebel, 1993; Boulton, 1994).

In the early 1990s, Kyrgyzstan's dysfunctional economy was further debilitated by the disruption in trade relations with other former Soviet republics and was characterized by a steep fall in output, worsening macro-economic imbalances, hyper-inflation, increasing shortages and widespread unemployment as the government attempted to implement its pro-free enterprise approach to development (Bassani, 1993). Social and economic indicators placed the Kyrgyz Republic in the range of low- to middle-income developing countries rather than with the industrial reforming states, with over half of the national output and employment coming from agriculture, despite the relatively small area of fertile land. Three-quarters of their food requirements were met by imports. These problems were compounded by severe environmental pollution and poor infrastructural conditions, including chronic underfunding of medical facilities, unsafe drinking water and poor sanitation systems, which have resulted in widespread infectious, nutritional

and reproductive health problems (UNICEF/WHO, 1992).

Kyrgyzstan's policy of adopting a market economy was supported by financial aid from a number of international organizations including the World Bank, the International Monetary Fund and the Asian Development Bank (Platt, 1994). The loans were used to import economic necessities such as oil, pharmaceutical products, telecommunications and transportation equipment (Lloyd, 1993). Despite significant progress with respect to market-oriented reforms, economic growth has been stymied by the lack of resources, the absence of processing infrastructure, inefficient state enterprises and widespread corruption in both the public and private sectors. By 1996, 70% of private enterprises were bankrupt because the market for their products had not been established, and 30% of GDP came from a black economy of independent, unregistered traders (Anderson, 1999). By 1998, there had been a significant upturn in the economy and IMF directors were describing the improved performance as an 'inspiration' to other emerging markets (Harun, 1999). More recently, they have commended Kyrgyzstan's macroeconomic policy performance, which has resulted in strong growth, low inflation and stability in the foreign exchange market. Whilst applauding the Republic's improved fiscal performance, they also acknowledge that resolute implementation of the new economic programme will be needed to enhance prospects for sustained growth and reductions in external debt, which is currently $1.5 billion (Europe-East, 2002).

Between 1996 and 2001, Kyrgyzstan received hard currency grants from the European Commission totalling €42.5 million. The funds have been used to support the country's balance of payments, to finance food imports and to promote a number of reforms in the agricultural and social sectors (BBC, 2002b; Interfax, 2002). Foreign direct investment has been encouraged by the policy of privatization and the implementation of a favourable tax regime, property purchase legislation (excluding land) and unlimited repatriation of profits for international businesses. As a result, it has managed to grow its foreign investment over the past decade in a range of sectors including tourism. The economy still gains much of its momentum from mining, but the government is attempting to become more export-oriented and tourism is beginning to play an increasingly important and strategic role in this transformation not simply because of its invisible export potential, but because national identity and image creation, which are so vital for positioning the Kyrgyz tourism product, are strategically important political issues for the Republic. This relationship between tourism and the nation-state is not unique; there is a natural affinity in the sense that both have a profound interest in promoting place uniqueness and differentiation. The particular circumstances of Kyrgyzstan's political and economic trajectory within a Central Asian context and the Republic's internal divisions mean, however, that tourism's role in positioning the state is critical.

Socio-cultural Issues

Central Asia's ethnically based republics were carved from Turkestan by the Soviets after the 1917 revolution in anticipation of reordering Central Asian society, and there was no correspondence between their arbitrary borders, drawn up in the 1920s and 1930s, and the region's ethnic and linguistic situation. As a result, many Uzbeks, Kyrgyz, Kazaks and others were left out of place (Anon, 1993). Not surprisingly, the politico-ethnic map of the region is extremely complex and many border regions are the subject of international dispute. In Kyrgyzstan, the Osh and Batken *oblasts* (regions) in the south comprise islands of Uzbek and Tajik territory and the city of Osh has a large population of ethnic Uzbeks. Further, the border with Tajikistan is a source of destabilization with respect to Islamic Renaissance Party sympathies in the Muslim strongholds of Osh, Batken and Jalal-Abad and also to the ousting of local press by Uzbek newspapers and magazines (BBC, 2002c). The Stalinist division of Central Asia also produced three external Kyrgyz enclaves, two in Uzbekistan and one in Tajikistan, which are characterized by similar tensions.

This international conflict is paralleled by hostilities within Kyrgyzstan between ethnic,

clan and religious groups at a number of different levels and geographic scales. Despite making considerable progress since 1991, the Republic is still struggling to find its political and cultural identity. The country was one of the most Russified of the former Soviet states and yet its remote location on the periphery of the former USSR was a key factor in the endurance of much of its pre-Soviet cultural heritage. This has experienced a renaissance since independence because of the heightened awareness of nationality and a growing sense of historic perspective. The religious affiliation of the Kyrgyz population is primarily Muslim (83%) with 14% Russian Orthodox (UNDP, 1999) but they are generally less dogmatic in their religious beliefs than their Tajik and Uzbek neighbours. This reflects the dominance of the nomadic lifestyle including the retention of many tribal traditions after their nominal conversion, although in the southern provinces, where 30% of the population are Uzbek, Islam does play a major role in people's lives and this was the case even during the Soviet era. By comparison, the northern region is more liberal and industrialized, with Slavs assuming a dominant position in the urban, industrial sector – the legacy of all-union ministries in Moscow managing the largest industrial enterprises in Kyrgyzstan. The concentration of Slavs has restricted ethnic Central Asian access to jobs, modern housing and social services and, understandably, this has increased the inter-regional tension within the Republic (Huskey, 1995). It is interesting to note that a recent survey showed that 64% of Kyrgyz felt that the north/south contradictions were the main destabilizing factor in society (UNDP, 1999).

The ethnic identity of the Kyrgyz is strongly linked to their language and traditions; both have been zealously guarded by national policy since independence. In the early 1990s, the Akayev government, under significant public pressure, aggressively pursued a policy of 'Kyrgyzification' of the non-native population including the introduction of Kyrgyz as the official language, thereby forcing the remaining European population to use it in most public situations. Substantial pressure from Moscow successfully reversed this policy and in 1996 parliament adopted Russian as an official state language alongside Kyrgyz. The large majority of the population can speak Russian and do so, especially in the north. Nevertheless, as in other Central Asian countries, there has been a significant emphasis on ethnic self-differentiation as a reaction against Soviet ethnic generalization and the stereotyping of non-Slavs.

The geographic separation of the Kyrgyz population has tended to reinforce conservatism in society, exemplified by the significance which has been placed on family and clan ties even during the Soviet period, but particularly since independence. Kyrgyz identity is closely aligned with membership in one of three clan groupings, a left wing or *sol* comprising seven clans in the north and west of the country and two southern clans, the right wing or *ong* which contains one clan, the *Adygine*, and the *ichkilik*, a group of many clans, some of which are not of Kyrgyz origin (Anderson, 2000). Clan membership plays an important role in public life and support for fellow clan members is particularly strong in northern provinces although regional or clan loyalties dominate life on both sides of the north/south divide. Rivalry among Kyrgyz clans has proved to be a significant barrier to the unification of Kyrgyzstan behind Akayev's programme of reform.

Inter-clan tension often spills over into the country's political arena and since March 2002, when violence erupted in the Jalal-Abad region triggered by the arrest and detention of Azimbek Beknazarov for criticizing the government's stance on the border dispute with China, Kyrgyzstan's image as one of the most stable republics in Central Asia has been severely damaged. Exasperation with government corruption, widespread poverty, failure to provide key services and the inadequate Muslim representation in government are widely considered to be key factors underlying the riots (Anon., 2002; BBC, 2002a). The Republic's unresolved social and economic problems, particularly the high population growth rate (between one third to one half of its population is under the age of 15), rising unemployment, low standard of living, competition for water and grazing resources and contention over land allocation have heightened its inter-ethnic strife. Given the critical importance of political stability for the development

and maintenance of a positive destination image (Hall and O'Sullivan, 1996), the domestic strife of a country where 48% of the population is living at or below the poverty line (NSCKR, 2001) is likely to have a strong impact on prospective tourist and investor predispositions.

The Development of Kyrgyzstan's Tourism Product

During the Soviet era, much of Kyrgyzstan was closed to international visitors because of the Republic's role in military development and this has left a significant legacy of health problems relating to the mining of uranium (Anon., 1994). Ironically, Kyrgyzstan was a destination for health and sports, serving the recreational needs of the USSR. This activity was centred on the health spas around the 600 km shoreline of Lake Ysyk-Kol (meaning 'warm lake' in Kyrgyz) – the world's second-largest mountain lake – situated in the north of the country. The lake was also home to a military research complex near Karakol at its eastern end, where the activities included the testing of high-precision torpedoes. Health tourism also developed around hot springs, for example the sanatorium at Jeti-Oghuz, 25 km west of Karakol.

Since the demise of the USSR, Kyrgyz health and sports tourism have declined in popularity but nevertheless retain their position as the largest market segments in terms of visitor numbers. Tourist activities are now concentrated in the mountains and along the route of the Great Silk Road. Silk Road heritage has emerged as a major focus for international tourism in Kyrgyzstan, as elsewhere in Central Asia, not least because of World Tourism Organization (WTO) promotion of this linear attraction to focus attention on the resources of Central Asia as a catalyst for economic development (WTO, 1997, 2001). Given the Republic's physical terrain, the global growth in the tourism market and the willingness of financial institutions to fund tourism projects, it is not surprising that Kyrgyzstan has recognized the strategic potential of tourism in political, economic and social terms. State policy makers have two overriding concerns. First, to increase the tourism product's competitiveness to raise the *per capita* levels of foreign exchange earnings. Second, to enhance regional development by creating economic opportunities in some of the most underdeveloped areas of the Republic (KSATS, 2001).

Countries are often marketed as one tourism product, but consist of many including regions, cities, packaged tours and themed or special interest experiences (Morgan and Pritchard, 2002). Ultimately, the product is defined by the consumer experience, but prior to visiting destinations, products are defined by the way in which their benefits are perceived by prospective consumers relative to competitive offers. This positioning of the product relates to segmentation (Mykletun *et al.*, 2001) although in Kyrgyzstan's case, this concept is only weakly developed; geographic and product segmentation bases are employed. The current product portfolio, as marketed by the main tour operators, is a range of tours featuring trekking, mountaineering, caving, skiing, hunting, botany, ornithology and cultural heritage featuring the Great Silk Road and religious history. The government's vision for the further development of tourism is outlined in a recent report, 'Development of the Tourism Sector of the Kyrgyz Republic Until 2010' (KSATS, 2001). The document identifies four main markets: the traditional 'health and recreational tourism', 'cultural tourism' based mainly on Great Silk Road heritage, 'adventure tourism and mountaineering' and 'ecological tourism'.

Whilst much of Kyrgyzstan's tourism development has hitherto focused on the country's natural environment, the Republic's cultural heritage is beginning to emerge as a significant feature of the tourism product in parallel with the tourism market's desire for 'cultural otherness'. Silk Road heritage has traditionally been a key element in the tourism portfolio, but other facets of Kyrgyz history and culture are now being harnessed to satisfy a number of important objectives, not least the creation of national identity and place brand equity. Cultural heritage has the potential to evoke an emotional attachment to a country, an important factor to the government of a divided republic like Kyrgyzstan. It is also a key issue with respect to creating a more meaningful image in the minds of prospective investors and consumers. However, Kyrgyz cultural heritage is a contentious issue. The awakening of clan, ethnic and religious identity, particularly marked in the south of the country over the 1990s, has continued into the millennium and constrained the development of a coherent image for the Republic. In an attempt to establish a shared identity, particular emphasis has been placed on those aspects of Kyrgyz heritage which are common to, and recognized by, inhabitants of both southern and northern regions. These elements are the *Manas epos* and pastoral nomadism and both have emerged as core elements of the cultural tourism product, the latter as *Jailoo* tourism.

The *Manas epos*, some 553,000 lines, is an acclaimed collection of epic tales which documents the formation and life of the Kyrgyz people and represents the highpoint of a widespread Central Asian oral culture. *Manas*, the principal hero of the *epos*, who successfully established a homeland for the Kyrgyz and defended it from enemy forces, has now become the pre-eminent cultural heritage icon. The significance of *Manas* was highlighted in 1995 when the Kyrgyz government spent an estimated US$8 million on celebrations for the *Manas epos* millennium event and UNESCO declared it the 'International Year of *Manas*' in recognition of its importance in human history. *Manas* symbolizes the spiritual unity of the Kyrgyz, their democratic principles and their independence (Guttman, 1999) and, as such, the legend is playing a key role in the construction of both Kyrgyz ethnic identity and the country's cultural tourism product.

Before Soviet collectivization, the Kyrgyz were pastoral nomads and this tradition has spawned *Jailoo* tourism. *Jailoo* (Kyrgyz for 'summer pasture') is a domestic tourism phenomenon that describes the seasonal practice of Kyrgyz families moving with their livestock to mountain pastures in May. This cultural heritage experience is now popular with international visitors who are able to stay with families in the traditional *bozu* (yurts) and enjoy authentic Kyrgyz hospitality. A joint venture between the Kyrgyz and Swiss governments, the 'Helvetas Agro' project, has focused on this 'Shepherd's Life' initiative and provided appropriate training for participants in an attempt to improve the quality of planning, marketing and tourism services in order to optimize economic returns. Cultural tourism connected with both the *Manas-* and *Jailoo-*related products has helped to spread economic and social benefits into some of the more remote areas of the country such as the Talas and Naryn regions, respectively.

The analysis of Kyrgyzstan's tourism product using a neo-classical model suggests that development has been constrained by a wide range of factors that reflect the political and economic context. These include the lack of investment capital (including foreign investment), poor infrastructure, an uneducated and unskilled workforce, minimal facilities including insufficient accommodation outside the major centres and inconsistent standards of service quality and hygiene. Additional constraints have been highlighted by BDO Consulting (1996), Khanna (1996), Eckford (1997) and Dudashvili (2000). These include the difficulty and high cost of internal access and the failure of the government to create favourable conditions for tourism development including the lack of incentives for foreign direct investment, bureaucratic problems including visa regulations and regional instability. Political unrest has also hampered the progress of the 'new silk road' (Koenig, 1997) – the European Union's 'Transport Corridor Europe-Caucasus-Asia' (Traceca) – which will improve European road and rail access to Kyrgyzstan. From a marketing perspective, there are a number of additional issues. The Republic has a weak image, an unfamiliar language and competition from more familiar, established destinations. Additionally, the government lacks overall product control, has a low budget for promotional expenditure and, as noted earlier, is constrained in its marketing efforts by the internal politics of destination branding relating to the marked north/south clan, ethnic, religious and political cleavages.

Positioning the Tourism Product

Kyrgyzstan's state-initiated tourism development is 'embedded' (Riley, 2000) within its political, economic and cultural context. In brand management terms, both the Kyrgyz tourism product and the Republic itself could be described as under-positioned because aware-ness of the destination in the world's major tourist-generating areas is poor and the country is indistinguishable from other Central Asian states. Until the early 1990s, the Soviet Central Asian countries were generally perceived by the outside world as a homogeneous region. Since independence, there has been some differentiation between the individual states with respect to their perceived international investment potential, but there has been little discrimination in Western tourism markets with the exception of the more discerning end of the consumer spectrum. Clearly, the development of an appropriate 'official' image of Kyrgyzstan is a critical issue given the primacy of destination images in the tourist decision-making process (Gartner, 1993; Court and Lupton, 1997). This is particularly important in Kyrgyzstan's case because market exposure to either actual experience of the Republic or 'organic' images, communicated predominantly through the mass media, has been limited.

Under-positioning has resulted from a number of factors *inter alia* the lack of a strong national identity and state image, the lack of capital and expertise, underdeveloped business partnerships, the range of political interests and regional instability. Moreover, whilst the WTO promotion of the Great Silk Road as a linear attraction across Central Asia (WTO, 1997) was important in raising awareness of Kyrgyzstan outside CIS countries and the former USSR, it did not differentiate effectively between the Republic's tourism product and those of its neighbours. Similarly, the emphasis on the Republic's natural resources in the KSATS' tourism promotional literature has highlighted its scenic beauty and related benefits, but has not distinguished Kyrgyzstan from other Central Asian countries or competitors further afield in the tourism market.

A statement in the Kyrgyz Constitution in 1993 highlighted Akeyev's intention to position the country away from other CIS states towards the west, on the basis of its democratic society and free market economy, to become 'Asia's Switzerland' (Haberstrok, 2000). In order to achieve this, the government differentiated the Republic negatively with respect to not being a Communist state, and positively in terms of both presenting a comparative image of the natural landscape and outlining the government's political and economic intentions. In the same year, Prime Minister Chyngyshev attempted to consolidate this position on his European tour by underlining three key dimensions. First, he emphasized the concept of Kyrgyzstan as a secular bulkhead against Islamic fundamentalism: the Republic had previously banned political parties based on religion and loaned border guards to neighbouring Tajikistan to prevent Afghan rebels from entering the country. Second, he reaffirmed Kyrgyzstan's commitment to becoming a free-market economy. Third, he highlighted the benefit of access to and from China across a long common border (Anon., 1993). Here, the Republic was differentiated on the basis of its non-Islamic status and by highlighting the country's economic and geographic benefits compared with those of other newly independent Central Asian states. In tourism development terms, the Republic's political and economic stability and its close proximity to the Chinese market are important issues.

The effectiveness of the government's strategy should be considered from a number of perspectives. At the state level, the comparison with Switzerland has been effective in creating a distinctive position, particularly in a regional context. This has proved to be beneficial from the perspective of international relations and foreign investment in the short term at least. However, on the domestic front, it has had the effect of increasing inter-regional tension because of the ethnic, religious and political differences between the north and south of the state. From a tourism perspective, the link with Switzerland's scenic beauty presents the marketplace with an immediate and powerful image of the Kyrgyz tourism product against a known frame of reference. The short-term benefits of promoting the natural landscape of the Republic and establishing the position of the product in this manner may, however, be outweighed by the problems of differentiating the state from Switzerland later in the product life cycle because of the strength of the initial image and the difficulty of changing market perceptions once they have been established.

Kyrgyzstan's tourism marketing continues to be dominated by images of mountains and lakes and is reflected in the tourism promotional strapline: 'land of sky-high mountains'

(KSATS, 2001). This is understandable given the Republic's comparative advantage in natural resources, but most, if not all countries highlight their physical beauty and since Kyrgyzstan's inadequate facilities and poor service cannot be used as a basis for differentiation, in the absence of alternative advantages, there is a high risk of product substitutability. An important step in establishing a destination's differential advantage is the definition of its core values (Morgan and Pritchard, 2002) but this is a complex issue for a politically divided country that struggles to reach agreement on a wide range of issues. Traditionally, the Kyrgyz were nomadic pastoralists and the herding economy still continues in many regions. Not surprisingly, mountains, horses and yurts are powerful symbols of national identity in both northern and southern regions. For example, the circular, crossed smoke hole at the top of the yurt is featured in the centre of the Kyrgyz national flag. Although these traditional lifestyle references are characteristically Kyrgyz and arguably capture the spirit of the country, they do not clearly distinguish the Republic from Central Asian neighbours because they share similar traditions. By comparison, the Manas legend is rapidly becoming part of the Kyrgyz self-image in terms of the way in which they prescriptively see themselves as historically constructed and culturally re-constructed. From a positioning perspective, this representation of Kyrgyzstan's values and characteristics has significant potential as a symbol of, and a vehicle for, both unification and differentiation of the Republic and its tourism product.

Whilst a belief and interest in the Manas legend is shared to some extent across other nations in Central Asia because it features in other Turkic epics (Musayev, 1994), Manas is associated with only the Kyrgyz national identity. Manas could therefore be used effectively to develop a personality for the place brand through the creation of a more romantic image of, and a spiritual attachment to, Kyrgyzstan (see Hallberg, 1995; Lury, 1998; Westwood et al., 1999). However, whilst there is some indication that Manas' symbolization of independence and unification has been used to good effect in the arena of domestic politics, the concept of strategically marketing

Manas, to differentiate the Republic's tourism product and gain a competitive advantage in the international marketplace, is as yet underdeveloped. This is surprising given the importance of *Manas* to the national psyche and the cultural bias of many of the brochures, but reflects the, as yet, limited international appeal of *Manas* outside Central Asia. The low-key approach is also linked with problems of access to Talas (the region most closely associated with *Manas*), the State Agency of Tourism and Sport's lack of funds, low income levels in the domestic tourism market and the priority given to targeting international tourism markets. As a result, the official promotional material continues to be dominated by tourism products based on the country's natural resources. The increasing exposure given to the Republic's cultural heritage, including *Manas*, is nevertheless beginning to redress the imbalance in the portfolio and highlight the huge potential of this resource for attracting the gaze of prospective tourists in the sizable Central Asian market – because *Manas* is an epic which transcends spatial boundaries (Bashiri, 1999).

Summary

Kyrgyzstan's transformation has been hampered by its remote location and mountainous terrain, outdated political structures, complex economic legacy, ethnic rivalries and social tensions. Until recently, this resulted in declining industrial output, increased inflation, rising unemployment and widespread poverty. The Republic is now beginning to emerge as a distinct socio-political entity, despite the heritage it shares with other newly independent Central Asian states: Islamic faith, Turkic language and culture, deeply rooted clan/tribal loyalties, a pre-Soviet nomadic lifestyle and Marxist–Leninist rule. However, recent events have destabilized the basis of the Republic's differentiation from its neighbours: its commitment to democracy, free market economics and standing against Islamic fundamentalism. As global tension mounts in the aftermath of 11 September, and its repercussions are felt in the Middle East and Central Asia, the challenge for Kyrgyzstan is to construct a national identity that is both socially and politically acceptable and ensure that its inter-regional enmity does not deteriorate into the kind of politics that are currently being played out in Tajikistan, a country whose religion- and clan-based divisions have hitherto been more extreme.

Over the 1990s, Kyrgyzstan's economic growth was driven mainly by agriculture, mining and construction. The government has now recognized tourism's potential to integrate the Republic into the global economy and this sector is beginning to make a useful contribution to foreign exchange earnings and the creation of employment opportunities. Furthermore, the development and positioning of the tourism product are providing opportunities to forge a national identity and secure political re-imaging objectives to support the country's economic and social development goals. The government must establish a clear, credible and sustainable position for the Republic in a tourism market characterized by increasing regional and international competition, product parity and substitutability. Much of the embryonic tourism development that has hitherto taken place has been piecemeal in character and the marketing of tourism products has been constrained and fragmentary. This has resulted in a number of disparate images that have hindered the development of a coherent position for Kyrgyzstan.

At present, the country lacks the requisite institutional and administrative capacity, investment capital, expertise, political stability and coordination at a number of spatial scales to underpin the development and promotion of tourism. Moreover, whilst it is acknowledged that the development of a viable tourism industry in Kyrgyzstan is a long-term process and that the political will exists to support this strategy, the Akayev government must address a number of significant constraints if it is to create an effective place brand, a concept that is only poorly developed at present. The weight of evidence suggests that, in tourism terms at least, the position of this least-known of the little-known former Soviet Asian republics is likely to remain marginal for the foreseeable future.

References

Anderson, J. (1999) *Kyrgyzstan: An Island of Democracy in Central Asia*. Harwood Academic Publishers, Amsterdam.

Anderson, J. (2000) Creating a framework for civil society in Kyrgyzstan. *Europe-Asia Studies* 52(1), 77–93.

Anon. (1993) On the map. *The Economist* 326 (7799), pp. 33–34.

Anon. (1994) Deadly secret: Kyrgyzstan. *The Economist* 333 (7892), pp. 45–47.

Anon. (2002) The shock of dissent: unrest in Central Asia. *The Economist* 23 March, pp. 31–32.

Baker, M.J. (1992) *Marketing Strategy and Management*, 2nd edn. Macmillan, Basingstoke.

Banks, J. and Ebel, R. (1993) Kyrgyzstan: problems and opportunities. *The Oil and Gas Journal* 91(11), 21–28.

Bashiri, I. (1999) *Manas: the Kyrgyz Epic*. University of Minnesota, Minneapolis, Minnesota <http://www.iles.umn.edu/faculty/bashiri/manas/manas.htm>.

Bassani, A. (1993) Steppes to a market economy. *OECD Observer* 180, 15–19.

BBC (2002a) Rivalry between Kyrgyz clans in echelons of power analysed. *BBC Monitoring International Reports*, Financial Times Information Ltd, 18 January.

BBC (2002b) European Commission to grant over 9 million euros to Kyrgyzstan for food security. *BBC Monitoring International Reports*, Financial Times Information Ltd, 11 April.

BBC (2002c) Uzbeks muscling in on Kyrgyz media market in the south. *BBC Monitoring International Reports*, Financial Times Information Ltd, 1 March.

BDO Consulting (1996) *Constraints to Tourism Development in Kyrgyzstan*. World Bank/Netherlands TA Trust for Central Asia, Amsterdam.

Boulton, L. (1994) Kyrgyzstan needs Russia. *Financial Times* 13 June, pp. 4–5.

Bruder, E.T. and Scanlan, K.A. (1993) Central Asia: a new business frontier. *Business America* 114, 2–6.

Coshall, J.T. (2000) Measurement of tourists' images: the repertory grid approach. *Journal of Travel Research* 39(1), 85–89.

Court, B. and Lupton, R.A. (1997) Customer portfolio development: modelling destination adopters, inactives and rejecters. *Journal of Travel Research* 36, 45–53.

Dudashvili, S. (2000) Tourism needs more than mountains. *The Bishkek Observer*, Bishkek, 1 October, p. 14.

Eckford, P.K. (1997) *International Tourism Potential in Ysyk-Kol Oblast, the Kyrgyz Republic: Report and Analysis*. World Tourism Organization, Madrid.

Europe-East (2002) IMF concludes 2001 consultation with the Kyrgyz Republic. *Europe-East* 23 January, pp. 317–318.

Finifter, A.W. (1996) Attitudes towards individual responsibility in post-Soviet societies. *American Political Science Review* 90, 138–145.

Fletcher, J.F. and Sergeyev, B. (2002) Islam and intolerance in Central Asia: the case of Kyrgyzstan. *Europe-Asia Studies* 54(2), 251–276.

Gartner, W.C. (1993) Image formation process. In: Uysal, M. and Fesenmaier, D. (eds) *Communication and Channel Systems in Tourism Marketing*. Haworth Press, New York, pp. 191–215.

Gibson, J.L. (1998) A sober second thought: an experiment in persuading Russians to tolerate. *American Journal of Political Science* 42(1), 819–850.

Gunn, C. (1974) *Vacationscape: Designing Tourist Regions*. Bureau of Business Research, University of Texas, Austin, Texas.

Guttman, C. (1999) Kyrgyzstan: breaking out of the old shell. *UNESCO Courier* November, pp. 21–22.

Haberstrok, M. (2000) *Kyrgyzstan – Asia's future Switzerland*. Comprehensive Development Framework, Bishkek <http://stat-gvc.bishkek-su/Eng/Home/Poymon.htm>.

Hall, C.M. and O'Sullivan, V. (1996) Tourism, political instability and violence. In: Pizam, A. and Mansfield, Y. (eds) *Tourism, Crime and International Security Issues*. John Wiley & Sons, Chichester, UK, pp. 105–122.

Hall, D. (2002) Branding and national identity: the case of Central and Eastern Europe. In: Morgan, N., Pritchard, A. and Pride, R. (eds) *Destination Branding: Creating the Unique Destination Proposition*. Butterworth-Heinemann, Oxford, pp. 87–105.

Hallberg, G. (1995) *All Consumers Are Not Created Equal*. John Wiley & Sons, New York.

Harun, J. (1999) Evolution of a Central Asian economic dynamo. *Business Times (Malaysia)* 31 August, pp. 6–7.

Hiro, D. (1993) Central Asian Republics retain the hierarchy of the Communist Party. *New Statesman and Society* 8 January, pp. 19–21.

Hitchcock, M., King, V. and Parnwell, M. (eds) (1993) *Tourism in South-East Asia*. Routledge, London.

Huskey, E. (1995) The rise of contested politics in Central Asia: elections in Kyrgyzstan, 1989–90. *Europe-Asia Studies* 47(5), 813–833.

Huskey, E. (1997) Kyrgyzstan: the politics of demographic and economic frustration. In: Bremner, I. and Taras, R. (eds) *New States, New Times: Building the Post-Soviet Nations*. Cambridge University Press, Cambridge, pp. 654–676.

Interfax (2002) New Agency, Moscow <http://web2.infotrac.galegroup.com/itw>.

Khanna, M.K. (1996) *Potential for Attracting More Indian Tourists to Kyrgyzstan*. Government of India, New Delhi.

Koenig, R. (1997) New silk road hits obstacles. *Journal of Commerce* 23 June, pp. 1–2.

KSATS (2001) *Development of the Tourism Sector of the Kyrgyz Republic Until 2010*. Kyrgyz State Agency of Tourism and Sport, Bishkek.

Linz, J. and Stepan, A. (1996) *Problems of Democratic Transition and Consolidation*. Johns Hopkins University Press, Baltimore, Maryland.

Lipset, S.M. (1994) The social requisites of democracy reviewed. *American Sociological Review* 59(1), 1–22.

Lloyd, J. (1993) IMF watches as Kyrgyzstan fights the battle of the Som. *Financial Times* 21 May, pp. 5–6.

Lury, G. (1998) *Brandwatching*. Blackhall, New York.

Miller, A.H., Hesli, V.L. and Reisinger, W.M. (1994) Reassessing the support for political and economic change in the former USSR. *Political Science Review* 88, 399–411.

Morgan, N. and Pritchard, A. (2002) Contextualising destination branding. In: Morgan, N., Pritchard, A. and Pride, R. (eds) *Destination Branding: Creating the Unique Destination Proposition*. Butterworth-Heinemann, Oxford, UK, pp. 11–41.

Musayev, S. (1994) *The Epos Manas*. Kyrgyz Polygraph Kombinat, Bishkek.

Mykletun, R.J., Crotts, J.C. and Mykletun, A. (2001) Positioning an island destination in the peripheral area of the Baltics: a flexible approach to market segmentation. *Tourism Management* 22, 493–500.

NSCKR (2001) *Report of the National Statistical Committee of the Republic of Kyrgyzstan*, NSCKR, Bishkek <http://stat-gvc.bishkek-su/Eng/Home/Poymon.htm>.

NSCKR (2002) *National Statistical Committee of the Republic of Kyrgyzstan*, NSCKR, Bishkek <http://stat-gvc.bishkek-su/Eng/Home/Poymon.htm>.

Nellis, J. (1999) Time to rethink privatization in transition economies? *Finance and Development* 36(2), 16–20.

Platt, G. (1994) Development bank approves financing for Kyrgyzstan. *Journal of Commerce* 14 September, pp. 3–4.

Riley, R.C. (2000) Embeddedness and the tourism industry in the Polish Southern Uplands: social processes as an explanatory framework. *European Urban and Regional Studies* 7, 195–210.

Starr, F. (1996) Making Eurasia stable. *Foreign Affairs* 74(1), 80–93.

Tapachai, N. and Waryszak, R. (2000) An examination of the role of beneficial image in tourist destination selection. *Journal of Travel Research* 39(1), 37–44.

UNDP (1999) *Kyrgyzstan: National Report on Human Development*. UNDP, Bishkek.

UNICEF/WHO (1992) *The looming crisis and fresh opportunity: health in Kazakhstan, Kyrgyzstan, Tajikistan, Turkmenistan and Uzbekhistan*. UNICEF/World Health Organization, New York.

Westwood, S., Morgan, N.J., Pritchard, A. and Ineson, E. (1999) Branding the package holiday: the role and significance of brands for UK tour operators. *Journal of Vacation Marketing* 5(3), 238–252.

Williams, A.M. and Shaw, G. (eds) (1998) *Tourism and Economic Development*, 3rd edn. John Wiley & Sons, Chichester, UK.

Wood, R. (1997) Tourism and the state: ethnic options and constructions of otherness. In: Picard, M. and Woods, R. (eds) *Tourism, Ethnicity and the State in Asian and Pacific Societies*. University of Hawaii Press, Honolulu, Hawaii, pp. 1–34.

WTO (1997) *WTO Reopens the Ancient Silk Road*. World Tourism Organization Press Release, Istanbul.

WTO (2001) *Study on Visa Facilitation in the Silk Road Countries*. World Tourism Organization, Madrid.

9   Transition for EU accession: the case of Malta's restructuring tourism and transport sectors

Maria Attard and Derek Hall

Introduction

The micro-state of Malta comprises a group of three islands situated 100 km south of Italy in the central Mediterranean. Malta, the largest island, is 246 sq km in area, Gozo is 67 sq km and Comino just 2 sq km. The total population was 391,415 in 2000, of whom 28,000 live in Gozo and Comino. With one of the highest densities of population and car ownership in Europe, Malta faces problems of land, space, and increasing transport pressures. Malta's major economic activity is tourism, which is highest during the summer months, placing further impacts on this small land mass. Although endowed with a rich history and cultural heritage, Malta's lingering image is a sea and sun destination catering to the mass, especially UK, market. In the process of trying to change that image, and to attract higher-value niche tourists to the island, a wide range of particularly environmental conflicts has arisen (Markwick, 2000, 2001; Bramwell, 2003). Increase in the provision of 5-star hotels has seen major land-take and heightened congestion at cultural sites. Resultant patterns of uneven regional development are not untypical for the Mediterranean (Hadjimichalis, 1987; Dunford, 1997). Malta's path of accession to the European Union has acted as a catalyst for change in many aspects of national policy. Figure 9.1 presents the distribution of tourist accommodation and resorts and some of the most popular cultural attractions.

The aim of this chapter is to examine the relationships between tourism, transport and Malta's approach to EU membership. There is a strong relationship between tourism and transport and it is evident that tourism would benefit from stricter EU regulations in the transport and environment sector to ensure improved service and overall improved tourism product. There are, however, conflicts in the direction the government is taking in the promotion of both tourism and the development of a national transport policy – aimed at reducing private car use and encouraging the use of the extensive public transport services.

Malta and the European Union

Malta, a former British colony, gained independence in 1964 and was declared a republic within the British Commonwealth in 1974. Until the 1960s the Maltese economy depended mostly on the British armed services and the naval dockyard (Mizzi, 1994). On independence, due to the islands' small size, their insularity from major markets and limited range of natural resources, development options appeared severely limited (Bramwell, 2003). Subsequently, light industry (textiles, electronic components) and tourism were

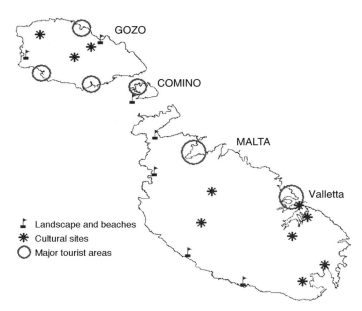

Fig. 9.1. The Maltese Islands – distribution of tourist areas and cultural sites (source: compiled by Maria Attard).

developed as major sources of employment (King, 1979; Lockhart and Mason, 1989; MTA, 1999). Three years after its independence, Malta informed the European Economic Community (EEC) that it wished to establish formal contractual relations. Subsequent negotiations led to the signing of the EC–Malta Association Agreement in December 1970. The Malta agreement came into force in April 1971, and covered trade-related issues, legal approximation and other areas of cooperation, including industry, environment, transport and customs (EC, 2000). It formalized close links between the Community and Malta and in particular the gradual removal of trade barriers to allow unhindered access to each other's markets. The result has been close trade relations, with the EU accounting for more than a half of Malta's exports and over two-thirds of total imports (EC, 1999a).

Since independence, Malta's essentially two-party political system has generated a fierce political debate within the islands but has also been severely divisive. Indeed, Cini (2002) regards Malta as having one of the purest two-party systems in the developed world, a characteristic particularly ironic given that it is a small and relatively ethnically homo-geneous state. The resulting intense political rivalry and polarization of Maltese society is complemented by a culture of patronage, exacerbated by the fact that elections have often been won by small margins of votes, which has often led to inaction due to an inability to pass and enforce laws. As a result, government policies 'may end up as a tattered patchwork of conflicting client demands' (Mallia, 1994: 700). Within this clientelistic context, political polarization can also be viewed in terms of 'traditionalists' and 'modernizers' (Mitchell, 2002): those who wish to see Malta moving forward via a process of Europeanization, and those who prefer to cling to 'Maltese ways of doing things'.

This division has complicated Malta's path to EU accession. Cini (2000) has argued that the process of 'Europeanization' can be defined in two ways: (a) as a process of adaptation to European rules and norms, and (b) as a redefinition or restatement of national identity. She suggests that while Malta's engagement in this process certainly coheres with the first of these dimensions, in the second case, the co-existence of two conflicting versions of national identity contributes to the party political and social divide that exists between those supporting EU membership and those opposing it. In this latter sense there remains a degree of ambivalence surrounding the 'Europeanization' of Malta. Baldacchino (2002) contends that this situation is partly the result of Malta's status as a 'nationless state', now characterized by post-colonial, small island nationalism, but with a history of attempted integration with France, Italy, Britain and Libya. However, Pace (2002) suggests that Malta's small size, although potentially an obstacle to its negotiating position with regard to EU entry, may also allow the Maltese to win concessions from the EU.

In July 1990, with the Nationalist Party in power, the government of Malta formally applied for European Community membership. The publication of a favourable opinion by the European Commission in June 1993 added momentum to this process, and EU–Malta relations became oriented towards the goal of accession.

With the coming to power of a new (Labour) government in October 1996, Malta's application for EU membership was placed in abeyance, but a return to government of the Nationalist Party in September 1998 saw accession negotiations recommencing. In February 1999 the European Commission recommended that the European Council give the go-ahead for the screening of Malta's legislation with a view to enabling the start of negotiations by the end of 1999. However, the Maltese press reported that Brussels was concerned that the anti-EU membership lobby – led by the opposition Labour Party – appeared to be more organized and vocal than its counterpart on the pro-membership side, and that the Maltese government needed to organize a distinctive and clear policy in favour of EU membership as other applicant countries had done (Manduca, 1999).

The European Commission's 1999 *Regular Report* (EC, 1999b) concluded that Malta was a functioning market economy and should be able to cope with competitive pressures and market forces within the EU, provided it continued with industrial restructuring (Hall, 2000). The EU leaders' meeting in Helsinki in December 1999 formally invited Malta and five other applicants to start accession talks in February 2000, which began subsequently.

The first Accession Partnership for Malta was decided in March 2000. Commission President Romano Prodi indicated that he considered Malta to be one of the most well prepared candidates (MFA, 1999). Negotiations with the EU were closed in December 2002 and Malta held a referendum in March 2003. A 53.65% majority vote in favour of membership led the government to call an early general election in April. The governing Nationalist Party won 51.7% of the vote based on a 96% turnout, reinforcing the government's confidence in signing acceptance of the package presented to Malta by the European Union for entry in May 2004.

It was during this process that increasing attention was paid to Malta's growing traffic problem, the incompatibility with EU accession of its archaic public transport system, their implications for the vitality of the tourism industry and to ways of addressing these issues. This was reinforced by the fact that the EC's *Regular Report* for 2002 (CEC, 2002: 60–62, 84–88) highlighted the 'very limited' progress that Malta had achieved since the previous report in the field of road transport, and highlighted the need for administrative capacity strengthening in both the transport and environment fields.

Restructuring and Re-imaging the Tourism Industry

Malta lies in a central position in the Mediterranean basin, and as an island it has the typical Mediterranean climate with hot, dry summers and cool, wet winters. Particularly since independence, Malta has come to rely more heavily on tourism as the major industries related to shipbuilding and port activities have declined. In 2001, 25% of Malta's GNP was derived from tourism, and the industry employed 9321 people (Micallef Grimaud, 2001a). The density of tourist accommodation is the highest in the Mediterranean, with a range of 1- to 5-star hotels mostly distributed in northern and eastern coastal areas (Table 9.1).

Malta has attracted more than a million tourists every year since the early 1990s. Until the mid-1980s, resort-based mass tourism, largely from the UK market, reinforced Malta's

Table 9.1. Malta: number and type of hotel establishments.

Type of hotel	Number of establishments
1-star hotels	5
2-star hotels	25
3-star hotels	45
4-star hotels	37
5-star hotels	11

Source: Malta Tourism Authority, 2001.

role as a sun, sand and sea destination, with the concentration of tourist arrivals in summer and early autumn (Fig. 9.4). This seasonality has heightened pressures on the country's infrastructure, especially energy supply, water quality, waste disposal and transport provision (Lockhart, 1997). It has also generated mixed responses from the Maltese themselves (Secretariat for Tourism, 1993; Planning Authority, 1997b; MTA, 2000a; Bramwell, 2003). However, this factor is complicated by:

- the increasing incidence of retirement in-migration, particularly from the UK, with migrants being both participants in, and recipients of, visiting friends and relatives tourism (Williams *et al.*, 2000);

- the difficulty of identifying tourism as an agent of change amongst a range of 'modernization' factors;
- new housing appearing to have contributed much more to land-take than has new tourism development (Ministry of Tourism, 1999; Planning Authority, 2001);
- the further complication that an unquantified number of new apartments and houses are used by Maltese as second homes, by returning emigrés, by international tourists and by retirement in-migrants.

Certainly, as tourism has grown, it is clear that the resulting increased physical congestion in public transport, road traffic and on beaches has become less acceptable (Boissevain and Theuma, 1998).

The country's 1989 tourism masterplan, based upon a consultants' report (Horwath and Horwath, 1989), called for the development of higher-quality tourism and for the encouragement of winter and spring attractions to reduce the high level of seasonality of international arrivals. As a result, sports such as diving and golf, alongside cultural and rural tourism, were given a higher level of development encouragement and promotion (Markwick, 1999, 2000). At the same time, the

Fig. 9.2. Malta: coastal hotel and apartment developments and wartime relic.

Fig. 9.3. Malta: the attractive harbour at Marsaxlokk.

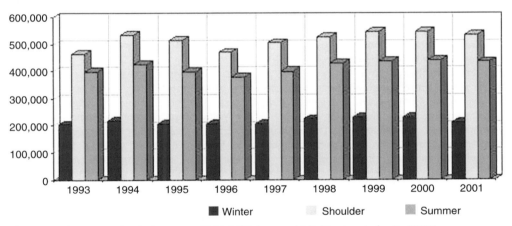

Fig. 9.4. Malta: seasonal arrivals pattern, 1993–2001 (source: Malta Tourism Authority, 2002).

government discouraged poor-quality development and prohibited the building of hotels of less than 4-star grading (Markwick, 2001). Such a policy has, however, led to environmental conflicts involving both the natural resources and residents of the island. On the one hand, the promotion of golf requires the diversion of scarce water resources and valuable land-take (Planning Authority, 1997a; Markwick, 2000). On the other, both logistical and moral conflicts have arisen as a consequence of foreign tourists visiting the country-side and being intimidated by local hunters and trappers, of which there are some 20,000 (Fenech, 1992), especially where the latter have carried out their pursuits in 'protected' areas (Markwick, 2001). Attempts to align Malta's environmental legislation with EU requirements have precipitated strong reaction from the politically articulate hunting lobby on a number of occasions (Borg, 1996). These issues have entered party politics and have contributed to the often sharp tensions on questions of EU accession.

Restructuring of tourism administration saw the establishment of the Malta Tourism Authority in 1999, with the aim of 'advancing the economic and social activity of tourism in the national interest, by working with all stakeholders to develop a sustainable industry for current and future generations' (MTA, 1999). A new brand image was promoted (MTA, 2000b).

The strategic plan for 2002–2004 re-emphasized the need to:

- spread the arrival of tourists more evenly throughout the year to reduce pressure in peak months;
- provide effective product delivery and commitment to service quality; and
- increase efforts to establish a network of stakeholder alliances where motivation, co-ordination and a sense of common achievement is promoted (MTA, 2002).

The Tourism Authority has attempted to improve Malta's outward image in a number of ways:

- a reinvigorated promotional campaign in major markets: UK, Germany (Fig. 9.5), Italy and Hungary;
- the setting up of a lively interactive website (http://www.visitmalta.com) and several new tourist offices around the islands;
- employment of eye-catching promotional straplines such as 'the heart of the Mediterranean' (MTA, 2003); and

- placing a re-emphasis upon niche marketing, notably conference tourism, cultural tourism and golf.

Yet such a policy has continued to exert social and environmental pressures in terms of:

- traffic congestion pinch-points;
- tourist congestion and intrusion at cultural sites;
- excessive land-take at key locations for luxury hotels;
- land and water demands of golf courses;
- space demands and marine pollution impacts of marina development (Boissevain, 1994a,b, 1996a,b; Boissevain and Theuma, 1998).

Critically, this continued encroachment on the Maltese people's own leisure space echoes colonial times when the British administration prevented public access to large areas of rural Malta (Boissevain and Theuma, 1998: 114–115).

Public Transport Provision

Most tourists visiting the island use public transport as their main mode, as it is easy and cheap to use. The Public Transport Directorate within the Malta Transport Authority estimates that 27% of all bus users are visitors. The buses are considered by many to be part of Malta's heritage, and some date back to the mid-20th century (Clegg,

Fig. 9.5. Malta Tourism Authority promotion in Germany (source: Malta Tourism Authority).

1981) (Fig. 9.6). Occasionally tourists recount their experiences of using local buses in the pages of Malta's newspapers. Others have created websites which display contemporary and historical photographs and information about individual buses, the main bus terminus in Valletta, and even examples of bus tickets (Edwards, 2001). However, this working dimension of Malta's heritage will soon be removed – most of the old bus fleet is being replaced with new standard passenger vehicles in order to increase passenger comfort and safety and reduce atmospheric emission pollution.

Presently, there are 508 buses which work on alternate days. Only 250 buses are actually needed to run the current bus network. Buses are owner-driven and many of the alterations made to the exterior body are associated with the individuals, for example a Brincat-bodied 1948 Thornycroft Sturdy (Lidstone, 2001b). Fifteen per cent of all existing buses are regarded as unsafe (estimate quoted in a debate in parliament, 9 May 2000). The major problem has been the financing of the new buses (Page, 1999). Debates surrounding this issue, together with the setting up of a new ticketing system, began in the mid 1990s. In 2001, the Maltese Government offered a subsidy for a limited replacement of the buses. Only seven had actually been bought by December 2002, following agreement to subsidize 147 new vehicles at €60,000 each. The new buses conform to low-floor, EURO III engine design, and general EU standards and regulations on the carriage of passengers. However, during 2003, arrivals of new vehicles began increasing: 87 would be arriving from the Chinese manufacturer King Long, making China the source of 59% of planned new vehicles, while 40 Turkish-built BMC Falcons were expected alongside batches of vehicles built in Poland, Macedonia and Greece (Anon., 2003a,c). The Government has also relaxed second-hand bus importation regulations. The latest developments in vehicle replacement are being promoted by the government within Malta's bid to join the European Union (Lidstone, 2001a).

The network of regular urban and suburban bus services is extensive; around 75% of the population lives within 10 minutes' walk of a bus stop. There are 91 routes operated at an average frequency of 20 minutes (3700 daily trips performed). Every town and village has a direct bus service connection with the capital Valletta, and since the early 1990s, a number of additional 'direct' services, largely catering for the travel needs of tourists, have been introduced between resorts and the beaches or places of cultural interest.

Fig. 9.6. Malta: Bedford service buses from the 1940s and 1960s at Valletta.

Malta's public transport patronage, however, has been decreasing over the past decade. Figures published by the Public Transport Directorate (the former Public Transport Authority) show a 10 million-passenger loss over the last decade. Figure 9.7 shows the ticket sales for the Public Transport Association (Malta island only) excluding the sales of the number of day passes. This loss has occurred due to several factors: the increase in private vehicle ownership and car use – by 1999 70% of the population used a car as their main mode of transport compared to less than 55% in 1989 (Xuereb, 1999) – the ever-increasing number of minibuses offering private transport, the increase in private coach operators and the introduction of day passes for which there are no data available on public transport usage (V. Spiteri, 2002, personal communication).

The Public Transport Association (PTA), which is made up of 400 bus owners and drivers, works to protect the interests of bus drivers and their families. The PTA sets the driver and route shifts, collects all earnings and distributes them equally between drivers and calculates the subsidy, which is then provided by the Government. In this manner, each driver earns an equal amount of money. This public service obligation grants exclusive rights to current operators, and through the appropriate legal instruments ensures the strict prohibition of direct competition.

The Government, through the newly set up Malta Transport Authority, regulates the network of services by ensuring continuity of all routes and subsidized low fares. Private coaches and minibuses provide services to locals and tourists – these generally cater for tours, transfers and private functions – but according to the PTA, there are now too many providers of public transport for a decreasing market, both domestic and international (V. Spiteri, 2002, personal communication). Both locals and tourists consider taxis to be expensive; horse carriages are mostly located in the tourist areas. All major international rental companies are represented in Malta, together with a larger number of small local businesses that run rental vehicle services in tourist areas. Table 9.2 indicates increasing tourist use of hire cars by season, heightening pressure on infrastructure and the need for road space.

Overall, the number of motor vehicles on the islands has grown by 116% since 1985. The Household Travel Survey, prepared by the Malta Environment and Planning Authority (MEPA) in 1998, showed that there were four times as many work-related trips being made by car than by public transport (MEPA, 2002).

Malta now has one of the highest *per capita* levels of car ownership in the Mediterranean and Europe (Fig. 9.8), with over 500 cars per 1000 inhabitants (Malta Transport Authority, 2003b). The country has 2188 km of road, of which just 157 km are arterial roads with two

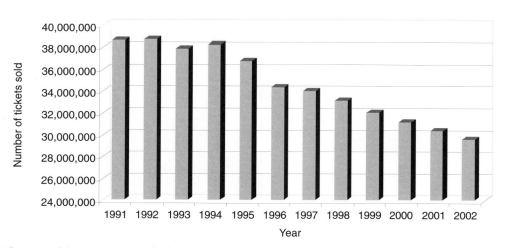

Fig. 9.7. Public transport annual ticket sales, 1991–2002 (source: Malta Transport Authority, 2003a).

Table 9.2. Malta: tourists' hire car use by season, 1998–2000.

Period	1998	1999	2000
Summer	129,000	132,000	133,000
Winter	68,000	68,700	69,000
Shoulder months	132,000	136,700	136,200

Source: Planning Authority, 1997.

or three lanes and 1275 km are urban roads (NSO, 2001a). Demand for road space is skewed by the very high density of population concentrated in just over 23% of the land area, exacerbated by a large tourist population and increasing dependence on the motor car. Table 9.3 presents a comparative overview with different European and Mediterranean countries.

Since 1994, in an effort to combat growing congestion problems in the main employment areas, a commuted parking payment scheme (CPPS) has been in operation whereby money is collected for new public car parks from developers who are unable to provide their own parking places. Extending this concept further, in 2003 the scheme purchased a num-

ber of low-floor electric buses to provide environmentally friendly park-and-ride routes (Anon., 2003a).

Impacts of EU Accession on Malta

Even though tourism is a major industry in Malta, accession to EU membership is unlikely to affect it directly. There are no specific chapters in the *acquis communautaire* (EU law) which specifically deal with the tourism industry. There are, however, a number of chapters which have a great influence on tourism. The most important of these relate to the environment, transport, agriculture and consumer affairs (Table 9.4).

In the transport chapter, some structural and legal changes will have an impact on tourism at many levels. Many of the issues deal with legal harmonization on the movement of goods, operators licensing for passenger and goods drivers, driver training and testing, and vehicle standards. There are also a number of requirements which will affect positively the state of the road network.

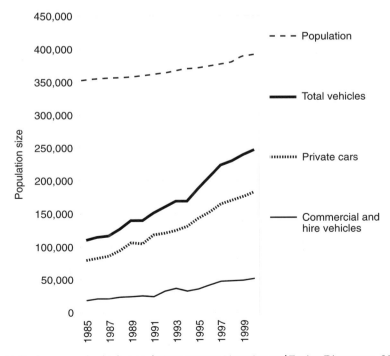

Fig. 9.8. Malta: human and vehicle populations (source: Licensing and Testing Directorate, 2002).

Table 9.3. Selected Mediterranean and European countries in comparison.

Country	Area (in km²)	Population density (per km²)	Roads (in km)	Rate of motorization[a]	Number of hotels and similar establishments	Tourist accommodation density[c]
Malta	316	1,234	2,188	501	248	0.78
Cyprus	9,251	82	10,654	340	580	0.06
Lebanon	10,452	383	6,359	312	:	:
Israel	22,145	266	15,464	212	:	:
Greece	132,000	80	470	275	8,168	0.06
UK	244,000	245	3,476[b]	414	51,292	0.21
Italy	301,000	192	6,621[b]	544	33,341	0.11
Spain	505,000	78	8,800[b]	424	16,229	0.03
France	544,000	108	11,000[b]	465	19,379	0.04
Turkey	814,578	77	382,000	63	1,862	0.002

Sources: Eurostat 2000, 2002.

[a] The rate of motorization is calculated as number of passenger cars per 1000 population.
[b] This figure refers to km of motorway.
[c] Establishments per km².

Table 9.4. European Community requirements in selected accession chapters.

Chapter	Description of community requirements
Environment	Improved water quality – nitrates, bathing water, drinking water, treatment of urban waste water (85% of Malta's urban waste water is discharged untreated into the sea) by producing a sewerage masterplan. Transposition of the habitats directive.
Transport	Mostly dealing with legal harmonization, VRT, speed limitation devices, road taxes, driving licences, air transport liberalization and removal of restrictions for cruise liners entering the Grand Harbour.
Agriculture	Problems with levies on local produce which will be removed, unattainable high productivity standards compared with EU, manage rural development, food safety.
Consumer and health protection	Concerns over consumer affairs, time-share and package travel, standardization of products.

Sources: Micallef Grimaud, 2001a; CEC, 2002.

The Transport Infrastructure Needs Assessment (TINA), carried out as part of Malta's commitment to the Trans-European Transport Network, has identified sections along the main road network for improvement and a need for upgrading the access road network. EU Structural and Cohesion Funds will support these network improvements upon accession at a cost of around €90 million. Many of the links identified for improvement are close to tourist areas or will make access to these zones much easier.

Public transport, primarily the regular urban and suburban public transport service, is probably the most affected sector in transport. The Malta Transport Authority has established stan-

dards to ensure compliance with EC regulations demanding the latest models available on the market (i.e. super low floor with EURO III engines). The full vehicle roadworthiness test (VRT) will be introduced by 2005, by which time it is hoped all current buses used for public transport will have been replaced.

It is estimated that only 40% of the 91 bus routes are, in fact, profitable. The PTA uses revenue from the profitable routes as cross-subsidy, but since 1995, the network of services has been operating at an overall loss: in 1999 revenue deficit subsidy amounted to €2,855,830 (Grech, 2001). This subsidy is complemented by:

- financial assistance in purchase of new buses;
- procurement of bus ticket-machine equipment for a national ticketing system;
- reduced levels of registration tax on new vehicle importation;
- operational assistance in dispatching buses; and
- provision of regular bus service infrastructure such as bus stops, bus stations and travel information (Malta Transport Authority, 2003b).

It is envisaged that the declining trend in patronage will necessitate continued state assistance in this sector. The subsidy will also remain after accession since it falls under public service obligation (D. Sutton, 2001, personal communication). Bus drivers will be affected by stricter regulations on driving licences and qualifications. Malta, being an island under 2300 sq km, has been exempted in national passenger transport operation from the application of EU legislation concerning maximum driving periods, minimum rest periods and break times.

In maritime transport, one of the major impacts will be the removal of restrictions on cruise liners entering Malta's Grand Harbour. This, coupled with the New Cruise Liner Terminal Project, which Malta has embarked upon, will increase the influx of cruise liners visiting the island. It is a growing sector in Malta, with a total of 127 ships in 1990 increasing to 360 ships in 2001 (MTA, 2001).

There will be positive and negative impacts with regards to air transport. Malta's carrier, Air Malta, operated a monopoly for several years, restricting other carriers and their fare rates. With liberalization Air Malta will have to compete on a European level with other carriers offering cheaper rates. Consequently, Air Malta will be able to operate in a more flexible manner, with access to more destinations. This should increase the number of international and domestic tourists using air transport. At 72 km from the nearest European mainland, Malta relies heavily on air transport. The total number of air passenger movements in 2000 amounted to 2.9 million, with Air Malta carrying 57% (NSO, 2001b).

Changes to transport legislation and policy, due both to the islands' development and the EU accession requirements, will benefit the resident population, the local economy and the tourism industry. Competition with European cities and other coastal areas will not be easy with limited resources. Better management and more proactive policy will be necessary to gain the optimum benefits of EU accession for transport and tourism.

Conclusion

As a central Mediterranean island(s) bastion (Lopasic, 2001), Malta has undergone considerable development over a relatively very short period of time. EU accession processes have stimulated an increased rate of development and change, to which policy making has had to respond. Since the re-opening of negotiations in 1998, the government has done its utmost to compensate for the 2-year loss suffered during the Labour Party administration, when the membership application was halted. A consultative referendum held in March 2003 showed a 53.6% 'yes' vote for Malta to join the EU, and general elections followed which consolidated this decision. However, the political process remained tense, with the opposition Labour Party insisting that Malta should not join the EU, while the Nationalist Party, currently in power, has worked hard towards the goal of accession since 1987.

These years of transition have seen a major shift in tourism product development. While mass tourism continues, policy focus now emphasizes high-value niche tourism. New brand images have been promoted in advertising and other media aimed at both the European market and EU opinion-formers, as an outward sign of Malta's modernization and acceptability within the European family. Yet social and environmental conflicts, exacerbated by the country's political polarization, have followed.

Malta's accession progress has seen major shifts in policy in a number of fields. Tourism will indirectly benefit from these changes and regulations imposed by the European Commission. Malta is restructuring its public transport system to better conform to EU standards. Other regulations imposed by the Commission will help to improve road safety through proper testing and licensing of passenger and goods transport.

Malta's domestic transport crisis particularly reflects increasing levels of car use, creating problems of pollution, congestion and landtake for parking facilities in prime historic and tourist areas. With increasing numbers of tourists arriving, coupled with the aim of attracting high-value niche tourists, it is to be expected that a shift in the balance of tourists' modal choice, from public transport to rental cars, will only exacerbate this situation. Government cooperation with the tourism and transport authorities is required to formulate strong policy to improve the product Malta is trying so hard to promote abroad.

References

Anon. (2003a) King Long wins Malta order. *Buses* 55(578), 10.

Anon. (2003b) Low-floor electric buses for Malta. *Buses* 55(579), 10.

Anon. (2003) New bus plan speeds up as Falcon lands in Malta. *Buses* 55(577), 11.

Attard, M. and Hall, D. (2003) Public transport modernisation and adjustment to EU accession requirements: the case of Malta's buses. *Journal of Transport Geography* 11(1), 13–24.

Baldacchino, G. (2002) A nationless state? Malta, national identity and the EU. *West European Politics* 25(4), 191–206.

Boissevain, J. (1994a) Discontent in Mdina. *Focus*, Summer, p.12.

Boissevain, J. (1994b) Cultural tourism development in Gozo: lessons from Malta. In: Briguglio, L. (ed.) *Tourism in Gozo: Policies, Prospects and Problems*. University of Malta, Msida, pp. 65–72.

Boissevain, J. (1996a) Ritual, tourism and cultural commoditization in Malta: culture by the pound? In: Selwyn, T. (ed.) *The Tourist Image: Myths and Myth Making in Tourism*. John Wiley & Sons, Chichester, UK, pp. 105–120.

Boissevain, J. (1996b) 'But we live here!' Perspectives on cultural tourism in Malta. In: Briguglio, L., Butler, R., Harrison, D. and Filho, W.L. (eds) *Sustainable Tourism in Islands and Small States: Case Studies*. Pinter, London, pp. 220–240.

Boissevain, J. and Theuma, N. (1998) Contested space: planners, tourists, developers and environmentalists in Malta. In: Abram, S. and Waldren, J. (eds) *Anthropological Perspectives on Local Development*. Routledge, London, pp. 96–119.

Borg, A. (1996) Rural versus urban: environmental perceptions in Malta. In: Fladmark, J. (ed.) *Sharing the Earth: Local Identity and Global Culture*. Donhead, Wimbledon, UK, pp. 111–125.

Bramwell, B. (2003) Maltese responses to tourism. *Annals of Tourism Research* 30(3), 581–605.

CEC (Commision of the European Communities) (2002) *2002 Regular Report on Malta's Progress Towards Accession*. CEC, Brussels <http://europe.eu.int/comm/enlargement/report2002/ml_en.pdf>.

Cini, M. (2000) The Europeanization of Malta: adaptation, identity and party politics. *South European Society and Politics* 5(2), 261–276.

Cini, M. (2002) A divided nation: polarization and the two-party system in Malta. *South European Society and Politics* 7(1) 6–23.

Clegg, W.P. (1981) Malta's unique bus fleet. In: Booth, G. (ed.) *Buses Annual 1982*. Ian Allan, London, pp. 40–45.

Dunford, M. (1997) Mediterranean economies: the dynamics of uneven development. In: King, R., Proudfoot, L. and Smith, B. (eds) *The Mediterranean Environment and Society*. Arnold, London, pp. 126–154.

EC (European Commission) (1999a) *Economic Developments and Structural Reform in the Candidate Countries. Country notes: Malta*. EC Economic Reform Monitor No. 3, Brussels <http://europa.eu.int/comm/economy_finance/document/docum_en.htm>.

EC (European Commission) (1999b) *Regular Report from the Commission on Progress Towards Accession: Malta – October 13, 1999*. EC, Brussels <http://europa.eu.int/comm/enlargement/malta/rep_10_99/x.htm>.

EC (European Commission) (2000) *Enlargement. Pre-Accession Strategy. Pre-Accession Instruments*. EC, Brussels <http://europa.eu.int/comm/energy_transport/en/1b_en.htm>.

Edwards, A. (2001) *Malta and Gozo Route Buses – the unofficial website* <http://website.lineone.net/~alan.c.edwards/maltabus.html#contents>.

Eurostat (2000) *Transport Trends in the Mediterranean Countries*. Eurostat, Luxembourg.

Eurostat (2002) *Eurostat Yearbook 2002*. Eurostat, Luxembourg.

Fenech, N. (1992) *Fatal Flight: the Maltese Obsession with Killing Birds*. Quiller Press, London.

Grech, B.J. (2001) *Government of Malta, Financial Report – 2000*. The Treasury, Valletta.

Hadjimichalis, C. (1987) *Uneven Development and Regionalism: State, Territory and Class in Southern Europe*. Croom Helm, Beckenham, UK.

Hall, D. (2000) Malta. In: Hall, D. and Danta, D. (eds) *Europe Goes East: EU Enlargement, Diversity and Uncertainty*. The Stationery Office, London, pp. 245–253.

Horwath and Horwath (1989) *The Maltese Islands Tourism Development Plan*. Horwath and Horwath (UK) Ltd, London.

King, R. (1979) Developments in the political and economic geography of Malta. *Tijdschrift voor Economische en Sociale Geografie* 70, 258–271.

Licensing and Testing Directorate (2002) *Transport Statistics*. Licensing and Testing Department, Valletta.

Lidstone, J.G. (2001a) Chinese whispers suggest the end may be nigh for 140 Maltese veterans. *Buses* 53 (558), 11.

Lidstone, J.G. (2001b) Malta prepares the ground for new low-floor buses. *Buses* 53(551), 11.

Lockhart, D. (1997) Tourism to Malta and Cyprus. In: Lockhart, D. and Drakakis-Smith, D. (eds) *Island Tourism: Trends and Prospects*. Pinter, London, pp. 152–178.

Lockhart, D. and Mason, K. (1989) *A Social and Economic Atlas of Malta and Gozo*. Department of Geography, University of Keele, Keele, UK.

Lopasic, A. (2001) Mediterranean islands: a concept. *Collegium Antropologicum* 25(1), 363–370.

Mallia, E. (1994) Land use: an account of environmental stewardship. In: Sultana, R. and Baldacchino, G. (eds) *Maltese Society: a Sociological Enquiry*. Mireva, Msida, pp. 685–705.

Malta Transport Authority (2003a) *Public Transport Ticket Sales*. Transport Strategy Directorate, Valletta.

Malta Transport Authority (2003b) Country Profile: Malta. Paper presented at the conference on *Public Transport Laws and Operating Systems*, Cagliari, Sardinia, 28 March.

Manduca, A. (1999) Brussels wants pro-EU lobby to be better organised. *The Malta Business Weekly*, 26 August <http://www.business-line.com/business-weekly/archives/253/15.html>.

Markwick, M. (1999) Malta's tourism industry since 1985: diversification, cultural tourism and issues of sustainability. *Scottish Geographical Journal* 115(1), 53–72.

Markwick, M. (2000) Golf tourism development, stakeholders, differing discourses and alternative agendas: the case of Malta. *Tourism Management* 21, 515–524.

Markwick, M. (2001) Alternative tourism: change, commodification and contestation of Malta's landscapes. *Geography* 86(3), 250–255.

MEPA (Malta Environment and Planning Authority) (2002) *Transport Topic Paper*. MEPA, Floriana.

MFA (Ministry of Foreign Affairs) (1999) The EU leaders meeting in Helsinki formally invited Malta and five other applicants to start accession talks in February 2000. *Foreign Affairs Diary* (Valletta), 6 December, pp. 1–3 <http://www.foreign.gov.mt/news/hron1.htm>.

Micallef Grimaud, J. (2001a) *2001 Regular Report on Malta's Progress Towards Accession: Implications for Tourism*. Ministry for Tourism, Valletta.

Micallef Grimaud, J. (2001b) *Report on Malta's Accession Negotiations with the European Union: 'What is the Outcome for Malta's Tourism Industry?'* Ministry for Tourism, Valletta.

Micallef Grimaud, J. (2002) *Framing the Future for European Tourism*. Ministry for Tourism, Valletta.

Ministry of Tourism (1999) *Carrying Capacity Assessment for Tourism in the Maltese Islands: Survey Draft*. Ministry of Tourism, Valletta.

Mitchell, J.P. (2002) Corruption and clientelism in a 'systemless system': the Europeanization of Maltese political culture. *European Society and Politics* 7(1), 43–62.

Mizzi, L. (1994) *Socio-economic Development and the Environment in Malta*. Centre for Mediterranean Studies, University of Bristol, Bristol, UK.

MTA (Malta Tourism Authority) (1999) *A New Brand Image for Malta*. MTA, Valletta <http://visitmalta. com/history.htm>.

MTA (Malta Tourism Authority) (2000a) *The Maltese Population's Perception of Tourism in Malta*. MTA, Valletta.

MTA (Malta Tourism Authority) (2000b) *The Malta Brand Image – Visual and Verbal Language*. MTA, Valletta <http://www.maltatourismauthority/dev/resources/corporate_man.shtml>.

MTA (Malta Tourism Authority) (2001) *Tourism Statistics 2001*. MTA, Valletta. <http://www.maltatourism authority.com/research/statistics>.

MTA (Malta Tourism Authority) (2002) *Malta Tourism Authority Strategic Plan 2002–2004*. MTA, Valletta <http://www.maltatourismauthority.com>.

MTA (Malta Tourism Authority) (2003) *Malta: Welcome to the Heart of the Mediterranean*. MTA, Valletta <http://www.visitmalta.com>.

National Statistics Office (NSO) (2001a) *Malta in Figures*. NSO, Valletta.

National Statistics Office (NSO) (2001b) *Transport Statistics 2001*. NSO, Valletta.

Pace, R. (2002) A small state and the European Union: Malta's EU accession experience. *South European Society and Politics* 7(1), 24–42.

Page, S. (1999) *Transport and Tourism*. Longman, Harlow, UK.

Planning Authority (1997a) *Golf Course Development in Malta: a Policy Paper*. Planning Authority, Floriana.

Planning Authority (1997b) *The Tourist Survey*. Planning Authority, Floriana.

Planning Authority (2001) *Structure Plan for the Maltese Islands: Tourism Topic Paper Draft for Public Consultation: Response to Comments*. Planning Authority, Floriana.

Secretariat for Tourism (1993) *Tourism and the Community: a Summary Report*. Secretariat for Tourism, Valletta.

Williams, A.M., King, R., Warnes, T. and Patterson, G. (2000) Tourism and international retirement migration: new forms of an old relationship in southern Europe. *Tourism Geographies* 2(1), 28–49.

Xuereb, K. (1999) Public transport use declines as private car ownership rises. *The Times* (Malta), 25 February. Reprinted by University of Malta at <http://www.um.edu.mt/pub/planauthsurv.html>.

10 Tourism development and planning in constrained circumstances: an institutional appraisal of the Turkish Republic of North Cyprus (TRNC)

Habib Alipour and Hasan Kilic

Introduction

For residents of developing countries, tourism has become a possible provider of, or catalyst for, economic growth or improvement (Gössling, 2000). In most island states the tourism industry has become the key vehicle for economic development and a major means of overcoming such structural weaknesses as a reliance on the export of one or two main products. Politically speaking, island states play a minor role on the world stage (Briguglio et al., 1996), but their overall structural capacities render them unique in character. Within such a perspective, this chapter evaluates the case of the Turkish Republic of North Cyprus (TRNC) in terms of the relationship between institutional structures and tourism development, and in comparison with processes which have moulded the tourism economy of Southern Cyprus. As part of the research process for this chapter, in-depth interviews were carried out with a number of people involved in the policy-making process and with plan implementation. The subjects who were interviewed were drawn from the National Tourism Organization (a public sector entity), the private sector (including travel agencies, hoteliers and tour operators), and organizations such as the media.

Cyprus and Tourism Development

Cyprus is the third largest island in the Mediterranean Sea. It occupies an area of 9851 sq km, 60 km south of the coast of Turkey, 96 km west of the coast of Syria and 322 km from Greece (Rustem, 1987) (see Fig. 10.1). Following the military intervention of Turkey in 1974 northern Cyprus was established as a separate entity, having a distinct political/administrative system covering about 37% of the island's territory (Andronikou, 1979: 237; Nedjatigil, 1982: 48–50). This partition has remained, peace being maintained by a United Nations presence. Since 1974 tourism has been developed extensively in the southern (Greek) portion of the island (Andronikou, 1987), but it has all but disappeared in the northern, Turkish portion, despite the fact that this area experienced most tourism development prior to partition (Lockhart, 1993; Butler and Mao, 1995).

A further important development in the evolution of the quasi-state of the North took place with a declaration of independence in November 1983. Following the raising of a petition by the people, the Turkish Federated State of Cyprus unanimously approved the establishment of the Turkish Republic of North Cyprus (TRNC), an entity recognized only by Turkey (Rustem, 1987; Frendo, 1993; Barkey

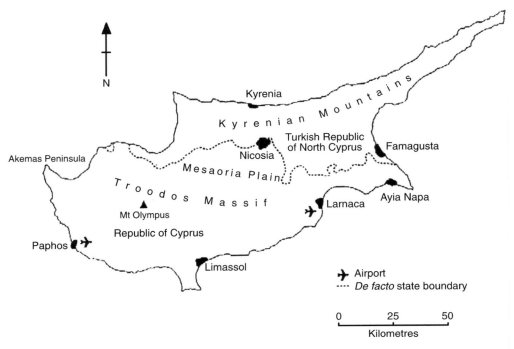

Fig. 10.1. Map of Cyprus with the demarcation line.

and Gordon, 2001). Despite this lack of recognition by the world community, northern Cyprus has diplomatic representation in a number of countries including the USA and Germany, and has been able to pursue trade relations with more than 60 countries (Nedjatigil, 1996: 11).

Partition of the Island/Partition of Tourism Development

Since 1974, tourism development has taken two distinct trajectories (Ioannides and Apostolopoulos, 1999: 51): one of rapid development in the South (Republic of Cyprus), and another of decline, underdevelopment and dependence in the North (Barkey and Gordon, 2001). Although covering just 37% of Cypriot territory, the Turkish sector inherited 50% of all catering and entertainment facilities, 82% of all accommodation and 96% of new hotels under construction (Andronikou, 1979), as well as 50% of the island's manufacturing capacity and two-thirds of its main agricultural area.

The case of the South (Republic of Cyprus)

Despite the interruption to tourism in 1974 (Ioannides, 1992), the southern Cyprus government's firm commitment rejuvenated the sector by formulating clearly defined policies. Coastal tourism was given priority attention and there was a conscious shift to higher-quality development. A comprehensive plan for tourism was prepared in the 1980s and new strategies covered product improvement, diversification and investment incentives. Provision of 10-year tax holidays for tourism enterprises undertaking large-scale product improvement projects was introduced (WTO, 1994). In terms of physical development, coastal areas were subjected to concentrated rather than dispersed development (Inskeep, 1991). Such a policy resulted in the former 'sleepy fishing village' of Ayia Napa accruing 34% of all tourist beds by the 1990s, the number having reached over 12,000. Indeed, 90% of all tourist beds in southern Cyprus can be found along its 295-km coastline (Godfrey, 1996; CTO, 1997; Lockhart, 1997). By 1986, almost 40% of the

31,883 registered bed spaces were in hotel apartments, compared to 29% of the total in 1981. In addition, there was a large non-licensed tourist accommodation sector – almost all self-catering – estimated to be 30,000 bed spaces (Ioannides, 1992: 722–723). The Cyprus Tourism Organization reported that of the 84,368 beds available at the end of 1996, 45,950 were in starred hotels, 26,715 in luxury 'A', 'B' and 'C' class hotel apartments, and the remaining 11,703 in tourist apartments, tourist villas, traditional houses, guest–houses and non-starred hotels.

The role of government in the evolution of the tourism sector and its rapid growth in South Cyprus is beyond dispute; with most of the credit going to the CTO, the South has consolidated its tourism development mainly as the result of domestic industry efforts rather than through international involvement, as is falsely perceived by the North. In the case of the Cypriot accommodation sector, the presence of foreign international firms is surprisingly small (Ioannides, 1992: 723). By the late 1990s, annual tourist arrivals exceeded 2 million, bed numbers exceeded 87,000 (Ioannides, 2000), and direct employment in the tourism industry stood at 40,000 (CTO, 1999). The tourism sector represented approximately 20% of the GNP, with receipts accounting for 43.7% of the total value of export of goods and services (CTO, 1999; Ioannides, 2000).

Tourism in the North (TRNC)

Since partition in 1974, tourism development in the North, in contrast to the South, has been characterized as either in decline or showing slow growth. This has been the case whilst the North's economy has remained highly dependent on foreign aid from Turkey (Barkey and Gordon, 2001). The North also suffers from diseconomies of scale in production, investment, consumption, education, and administrative services (Yorucu and Jackson, 1996: 2). The country has little industry, lacks water (e.g. see Hall, 2000), and has no mineral reserves. Meanwhile, in common with many other Mediterranean countries, tourism remains the main sector for potential economic development. Government tourism policies, however, lack clear strategic direction compared to those of the South. Despite inheriting 387 km of coastal resources and most of the tourism accommodation in Famagusta and Kyrenia, as well as 65% of the total bed capacity, and the main access point of Nicosia International Airport (Saveriades, 2000: 148), northern Cyprus has failed to capitalize on this inheritance through lack of clear policy formulation and plan implementation. For both external and internal reasons, tourism in the North remains highly dependent on Turkey: for financial support, for the majority of its tourist market and as a gateway to the rest of the world. As a result, bed occupancy rates have rarely exceeded 50% and tourism's contribution, as a proportion of GDP, has remained insignificant, and relatively static at around 3% (MSDPM, 1999) (Table 10.1). Agriculture, in particular citrus production, remains the backbone of the economy (Scott, 1997: 65).

After 1974, the Turkish Cypriot administration did not stress tourism development but placed heavy emphasis upon creating a robust agricultural sector to support the predominantly rural population (Ioannides, 2000: 124), but limited success in this sphere prompted officials to shift their attention to other sectors such as textiles and tourism. Despite its potential, however, the tourism industry has not progressed, as it continues to be affected by an international embargo on the one hand, and shortsightedness of policy makers on the other.

Bed capacity has remained below 10,000, with uncompetitive quality in all respects (interview with directors of Cyprus Turkish Travel Agents Union (KITSAB) and Cyprus Turkish Hoteliers Union (KITOB), 2001). The low occupancy rate, which has driven many hotel owners to the verge of bankruptcy, is complemented by limited employment in the tourism sector – just 3% of the total employment. Agriculture remains the second largest employer after the public sector (SPO, 1999). Most university graduates who are trained in tourism do not find employment in the North, and tend to look for work in Turkey or emigrate (Akis and Warner, 1994; Kibris Daily, 2001).

The accommodation sector has continued to suffer because of the short average length of

Table 10.1. Tourism and its impact on the TRNC's economy, 1990–1999.

Year	Value added in tourism sector (in US$ million)	% of GDP	Net tourism income (in US$ million)	The ratio of net tourism income to the trade balance
1990	160.9	2.3	224.8	71.1
1991	125.8	1.9	153.6	61.8
1992	178.2	2.5	175.1	55.3
1993	221.9	3.0	224.6	72.6
1994	235.0	3.3	172.9	74.1
1995	243.6	3.3	218.9	73.3
1996	210.9	2.8	175.6	70.8
1997	241.7	3.0	183.2	61.3
1998	249.5	3.0	186.0	55.2
1999[a]	273.9	3.1	195.3	55.7

[a] Estimated value.
Source: MSDPM (1999).

Fig. 10.2. Northern Cyprus: Gazimağusa/Famagusta from the town's Venetian walls, looking towards the sea.

stay – 4 to 5 nights compared to 11 in the South – and because of the low percentage of non-Turkish arrivals: in 2001 over 56% of the total arrivals came from mainland Turkey, many of whom were on short trips to gamble, as gambling tourism has become the dominant form of tourism due to closure of casinos in Turkey (Avci and Asikoglu, 1998; MTE, 2001).

National tourism organization

It was not until 1993 that the North's tourism sector was placed under the supervision of the Ministry of Tourism, Communications and Public Works, when a coalition government was inaugurated. Also at this time the Ministry of State and Deputy Prime Ministry (MSDPM)

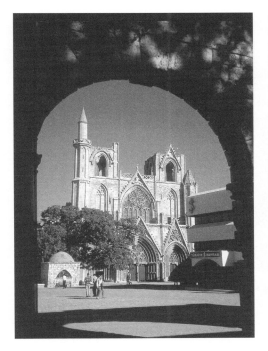

Fig. 10.3. Northern Cyprus: Gazimağusa/Famagusta: the Lala Mustafa Paşa Mosque, formerly the town's Cathedral of St Nicholas (1298–1326), modelled on Rheims cathedral.

was established as an agency responsible for the tourism sector. Yet MSDPM comes under the auspices of the Prime Minister's office and has no legal authority regarding tourism development. The policy-making process has to be delegated from the Prime Minister's office. This organizational structure renders MSDPM unable to take policy decisions. The lack of a separate tourism ministry is an indication of confusion in government over the tourism sector, and hesitation regarding a commitment to recognize its importance and significance. The lack of a statutory organization impedes efficient and effective public management, especially important in the implementation of tourism policy and marketing (Elliot, 1997; Inskeep, 1991). It also indicates a paradigmatic weakness of a public sector which has not been able to adjust to the evolutionary processes that are taking place globally in the tourism sector (Burns, 1999).

The first development plan (1983–1986)

The first 3-year development plan coincided with political upheavals which culminated in the declaration of independence (15 November 1983), and it was regarded as a plan for the

Fig. 10.4. Northern Cyprus: Salamis Bay, and a formerly Greek Cypriot-owned hotel complex.

Fig. 10.5. Northern Cyprus: part of the relatively untouched Karpas Peninsula.

transitional period (SPO, 1988). The plan lacked policy elaboration regarding the tourism sector, and was devoid of any plan target either at the national or local level. This was perhaps understandable given the government's preoccupation with the political upheavals in terms of the struggle for the declaration of independence on the one hand, and total economic dependency on Turkey on the other (Dodd, 1993). During this period tourism did not play a significant role in alleviating economic problems. An external trade deficit of US$101.2 million had amassed by the end of this plan period (Olgun, 1993). Receipts experienced an annual rise of 18% throughout the plan period, not related to an active tourist flow however, but rather to expenditure by the Turkish Peace Keeping Force and to income from the Free Zone (the UN based force) (SPO, 1988). The share of public sector investment in tourism experienced a decrease during this plan period. The sectoral contribution of tourism to GDP remained as low as 1.8% (SPO, 1988).

The second development plan (1988–1992)

Contrary to the first development plan, the second plan addressed the employment of

tourism as a means to achieve such policy objectives as:

- a balance of payments;
- improved exchange reserves;
- developing better links with European and Middle Eastern countries;
- an improved infrastructure;
- an improved quality of service; and
- raising the competitiveness of TRNC.

Overall aims were to increase the number of tourist arrivals as well as to alleviate the infrastructural bottlenecks which persisted as clear indications of fundamental flaws in the sector. The target of doubling income from tourism by the end of this plan period was never realized, while damaging and constricting infrastructural shortcomings continued (SPO, 1992).

The third development plan (1993–1998)

With reference to tourism, the third plan was critical of the lack of previous achievement, and it acknowledged:

- the absence of a physical masterplan;
- a lack of environmental concern due to the absence of appropriate institutions in this field;

- the lack of relevant laws and regulations;
- inattention to the safeguarding of historical resources; and
- an apathetic attitude towards the concept of national parks and conservation management (SPO, 1992). For example, the statutory designation of the Karpaz peninsula area of north-western Cyprus as a national park needs to be strengthened. The lack of common-sense laws and regulations, such as zoning and land use, leaves this valuable asset vulnerable to exploitation, as well as to its possible subjugation to rapid development in the event of a resolution to the Cyprus conflict being reached.

What went wrong?

An institutional approach

To evaluate more closely the case of the TRNC an institutional approach is adopted drawing on the work of North (1990, 1995) and others (Williams, 1998; Vandenberg, 2002) (Fig. 10.6). This recognizes a fundamental role for institutions in society as the underlying determinants of long-term economic performance (Aron, 2000: 1).

Institutionalist approaches draw on development theory (Sreekumar and Parayil, 2002; Theuns, 2002) and statist perspectives (World Bank, 1992, 1997; Barro, 1997), which argue

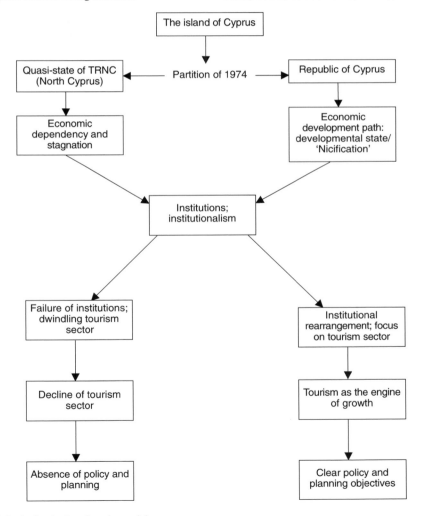

Fig. 10.6. An institutional study model.

that the key to economic growth is getting the sectoral composition of the economy right. An institutional approach also draws on social capital perspectives (Engelbert, 2000; see also Chapter 3) which view institutional quality as a function of a country's level of social capital.

At the heart of North Cyprus' underdevelopment lies an institutional matrix (Casey, 1998), which has remained too inert and vulnerable in its dependence on Turkey. Foreign investors have been reluctant to invest in an unstable country, with no independent currency system, exacerbated by a depreciating Turkish lira due to high inflation in mainland Turkey.

Governmental institutions in the North are subject to double stress, being both highly dependent on Turkey and subject to UN sanctions. Policy makers have failed to establish strong institutions and have fallen short of encouraging the formation of more appropriate administrative structures necessary for growth and development (Aron, 2000). In particular, government institutions have been unable to act as a real stimulus for private sector investment (Mansfeld and Kliot, 1996: 195). The weakness of political and economic institutions has been exacerbated by external elements, notably the embargo, which have played a major detrimental and contributory role in the stagnation of the economy as a whole, and of tourism in particular.

The role of institutions and structural deficiency

In the past decade research has increasingly focused on the ways in which institutions work to influence and shape their environments (Lawrence, 1999: 1). In the case of South Cyprus, despite the loss of major tourist resources to the North after partition, an institutional rearrangement with clear policy objectives played a decisive role in substantial economic growth with tourism as a major sector. This policy was characterized by government which increasingly saw itself as a 'developmental state' (World Bank, 1992), as the central provider of economic rules and their enforcement (Hira and Hira, 2000). Here the state acted as an independent institution

and a source of growth (Ioannides and Apostolopoulos, 1999; Ayres, 2000). This was also a parallel experience to that of a number of other smaller NICs such as Hong Kong and Singapore (Panayiotopoulos, 1995: 4). As such, the developmental state reconfigured institutions were:

> typified by an extensive planning apparatus which implemented policies to facilitate the expansion of the domestic private sector, in order to be able to address broader developmental concerns. A key characteristic amongst these issues was the centrality of tourism in the Cypriot economic development, which was on an even bigger scale, proportionally speaking, than the other Southern European NICs.
>
> (Panayiotopoulos, 1995: 4)

In terms of economic development, the emphasis was not just on the policies promulgated by the state, but also on the broader social, cultural, political and economic contexts in which those policies were implemented (Casey, 1998: 2). For example, concepts of 'social justice' and 'industrial peace' were adopted alongside 'growth', such that 'sacrifices' were made by Greek Cypriot organized labour in the national interest during the post-1974 period of reconstruction and development (Panayiotopoulos, 1995: 7).

In such a contextual framework, Dana and Dana (2000: 1) have explained the development of entrepreneurship as a function of differing beliefs, cultural values, government policies and ethnic composition. For example, in Cyprus both communities experienced a 'refugee' problem following partition. The manner in which this issue was dealt with demonstrated a marked contrast in policies. In the South, the policy was to redirect this problem into an asset for growth through the reorganization of the urban labour force, with the refugees providing new labour capital (Panayiotopoulos, 1995). By contrast, in the North, policies pertaining to the refugees transformed them into an 'idle' and largely unproductive labour force which became rent-seeking instead of becoming an innovative and productive workforce.

In the North, institutional failure that has constrained tourism development and growth has spotlighted the institution responsible for tourism. The National Tourism Organization in the North has failed to define itself in terms of statutory power (Inskeep, 1991) or decision-making capacity. The tribal character and kinship-oriented nature of public and private institutions have also impeded the ability of the tourism institution to furnish a strategy for the sector for the past two decades. Party politics have played a key role in the present structure of the tourism institution in terms of policy, strategy and planning. As each party comes to power, a new director is appointed to manage the tourism sector. The new manager has his/her party affiliation and kinship ties which inject shortsighted (Godfrey, 1996) organizational views, often closely linked to the socio-political structure of the destination. The implications of this process are a lack of opportunity to reflect on, formulate and design a comprehensive and strategic policy and plan for the sector (interview with the representatives of KITSAB and KITOB, 2001). It is not surprising, therefore, that for almost the last three decades, the idea of preparing a tourism masterplan has remained only as rhetoric (Lockhart, 1994). A lack of commitment combined with a considerable degree of apathy on the part of public sector management and politicians has resulted in the perpetuation of a superficial understanding of the sector.

A further fundamental institutional problem in the North has been the lack of a strategy commensurate with the circumstances that developed following partition. This resulted in the creation of a structural and legal vacuum (Lawrence, 1999). In the South, by contrast, institutional strategies were developed within the post-partition context to establish a strategically favourable set of conditions.

'Planning': the missing link?

The series of structural inadequacies that have handicapped tourism growth and development in northern Cyprus are transparent throughout the processes and stages of planning, policy and implementation (Inskeep, 1991; Gunn, 1994;), revealing deficiencies in relation to:

- a national development plan including tourism;
- a national plan establishing guidelines for the development of infrastructure;
- a national tourism development plan;
- a plan establishing guidelines for the development of tourism infrastructure at a national level; and
- a national plan for promoting and marketing tourist products at a national level (Wilkinson, 1997: 25).

The absence of a masterplan for tourism and the necessary laws to implement such a plan has resulted in the lack of a strategy for the sector which could enable it to aim for sustainable and competitive growth.

Recommended organizational model

Tourism development and planning, at least in most developing countries, has fallen within the public sector's domain. However, each country needs the type or form of organization or administration appropriate to its own socio-political and environmental characteristics (Inskeep, 1991: 411).

On this basis a hypothetical institutional model is proposed for tourism development in TRNC (Fig. 10.7).

A lack of institutionalization in the tourism sector has been the result of traditional bureaucratic approaches, and the subjugation of the tourism sector to a competition between the various public bodies seeking to enlarge their sphere of influence (Tosun, 2001). The formation of a body at cabinet level (a government ministry) with statutory power and with the ability to articulate, sponsor and defend particular practices pertaining to tourism, is of central importance.

Advisory board

An advisory board should be made up of public and private sector representatives who are either closely involved in tourism development, or who have a great interest in supporting the sector's rational development. The board's members may also be drawn from non-profit-making organizations. Members will also

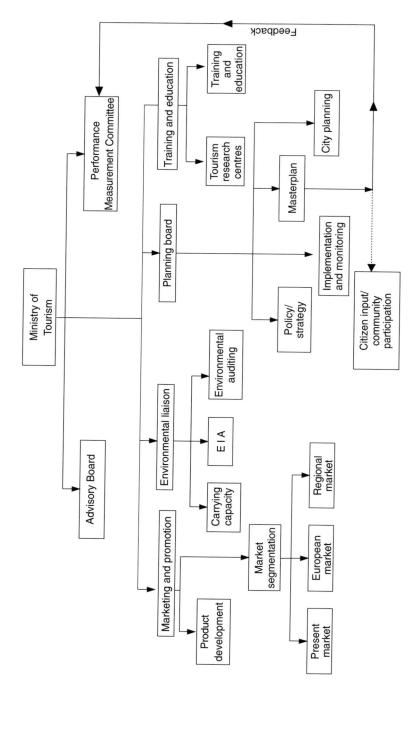

Fig. 10.7. A hypothetical national tourism organization structure for North Cyprus.

have acquired expertise by being in the sector and, therefore, will not be approaching the process in an *ad hoc* and short-term manner (Gunn, 1994). In the case of the TRNC, there is a range of representatives available to serve on such a board: the Cyprus Turkish Hoteliers Union, the Cyprus Turkish Travel Agency Union, representatives from conservation and environmentalist groups, as well as from the higher education establishments. The committee would be an independent body, which could become involved in policy making and offering advice, ensuring that strategy as laid out in the masterplan is followed. Institutionally, the committees would be able to investigate and take an in-depth look at potential future development, commission research, call for official papers, and collect evidence and interview witnesses (Elliot, 1997: 78). In the case of the TRNC, committees could establish a platform to prevent party politics from interfering with, and adversely affecting, the tourism sector. They should avoid political conflict and involve industry and the community to assist the policy process to be more transparent and democratic (Elliot, 1997: 78).

Performance measurement committee

The concept of an evaluation committee is taken from the experience of business corporations as well as from governmental practice. Such evaluation should ensure that the organization focuses on its strategic objectives (Schlegel, 1999: 1).

Departments

The model contains at least four departments that would be essential for establishing a structural framework. These are marketing and promotion, environmental liaison, a planning board, and training and education. Marketing and promotion consist of activities including product development and market targeting. Product development must provide a clear vision in relation to, and commensurate with, supply and demand (Gunn, 1994). Unfortunately, in the case of the TRNC, the lack of product diversification has led to vulnerability in the sector (Godfrey, 1996). South Cyprus'

initiative of 'rural tourism' development has been a response to a lack of product-led strategy, in recognition of the fact that rural tourism has become an important sector in European tourism (Sharpley and Sharpley, 1997; Roberts and Hall, 2001; Sharpley, 2003).

Environmental liaison would be a department designed to build an organic relationship with other environmental agencies in order to incorporate sustainability into the tourism planning process. What is currently missing is a strong environmental agency, environmental laws and policies as well as the implementation techniques required to achieve sustainability goals. The tourism sector should develop a new rapport with the currently weak environmental agency in order to strengthen the quality of the tourism product. Environmental impact assessment, environmental auditing, and carrying capacity standards are some examples of measures required for both the tourism sector and environmental organizations in the TRNC. The model suggests the building of a bridge between the environmental office and the tourism sector as a first step toward the notion of putting the 'eco' into tourism in the TRNC (Ivanko, 2001). Energy conservation can also be realized by the integration of environmental management with the Ministry of Tourism, which can, in turn, be linked to a performance measurement committee.

The current TRNC planning board requires revision and revitalization as it is suffering from a lack of both specialized staff and appropriate funds, as well as an overall lack of vision for tourism development. The model suggests that the board should be involved in the policy formulation process, the implementation of plans, and the monitoring process. It should also be in close contact and cooperation with the municipalities who are responsible for urban planning, development, and the periodic revision of the tourism masterplan. Unfortunately, planning in general, and tourism planning in particular, has not been sufficiently integrated to produce a nationwide agenda for growth and development. In relation to training and education, with the existence of several universities in the TRNC with training and education in tourism and hospitality alongside a government-supported tourism vocational school and publicly sponsored courses to train tour guides, the issue of skilled labour should not be an obstacle in the tourism sector. A main aim should be to develop collaborative links between the national tourism organization and educational institutions in tourism in order to ensure an appropriate supply of skilled labour.

Conclusion

This chapter has provided an overview of the development and structure of tourism in the Turkish Republic of North Cyprus since partition in 1974. A comparative analysis in relation to South Cyprus (the Republic of Cyprus) has been employed to demonstrate the different trajectories that tourism development has taken on the island. The chapter has also attempted to explore the institutional constraints on the tourism sector and its evolution, growth and dynamism.

In the case of the TRNC, major bottlenecks in the tourism sector *per se*, and constraints on overall economic growth, continue to reflect not only unfavourable external factors, but also a lack of institutional reorganization and adaptation to the new world environment. The tourism sector has remained a relatively low-key economic contributor despite its tremendous potential, being constrained by such factors as infrastructure weaknesses and an inadequate banking system (Akis and Warner, 1994).

Yet the North has remained relatively untouched and unspoiled in terms of its environment and landscapes. This provides current opportunities for innovation in the form of developing sustainable tourism, and for the avoidance of mistakes made in South Cyprus (Lockhart, 1997; Ioannides, 2000). There are various non-governmental organizations and concerned individuals who are willing to offer their expertise and assistance in devising and designing a new and productive policy in the form of a masterplan to ensure the positive redirection of the tourism sector in Northern Cyprus. Above all, governmental flexibility is required to permit and encourage cooperative working towards this end.

References

Akis, S. and Warner, J. (1994) A descriptive analysis of North Cyprus tourism. *Tourism Management* 15(5), 379–388.

Andronikou, A. (1979) Tourism in Cyprus. In: de Kadt, E. (ed.) *Tourism: Passport to Development*. Oxford University Press, London, pp. 237–264.

Andronikou, A. (1987) *Development of Tourism in Cyprus: Harmonization of Tourism with the Environment*. Cosmos, Nicosia.

Aron, J. (2000) Growth and institutions: a review of the evidence. *World Bank Research Observer* 15(1), 99–135.

Avci, T. and Asikoglu, S. (1998) Kumar turizminin kuzey Kibris Turk cumhuriyeti ekonomisine etkileri. *Anatolia* 9, 17–25.

Ayres, R. (2000) Tourism as a passport to development in small states: reflection on Cyprus. *International Journal of Social Economics* 27(2), 114–133.

Barkey, H.J. and Gordon, P.H. (2001) Cyprus: the predictable crisis. *National Interest* 66, 83–93.

Barro, R.J. (1997) *Determinants of Economic Growth: A Cross-Country Empirical Study*. MIT Press, Cambridge, Massachusetts.

Briguglio, L., Butler, R., Harrison, D. and Filho, W. (eds) (1996) *Sustainable Tourism in Islands and Small States: Case Studies*. Pinter, London.

Burns, P. (1999) Paradoxes in planning tourism: elitism or brutalism? *Annals of Tourism Research* 26(2), 329–348.

Butler, R.W. and Mao, B. (1995) Tourism between quasi-states. In: Butler, R.W. and Pearce, D. (eds) *Change in Tourism*. Routledge, London, pp. 92–113.

Casey, T.C. (1998) Economic policy, institutions, and economic growth in an era of globalisation. *Journal of Social, Political and Economic Studies* 23(4), 379–432.

CTO (Cyprus Tourism Organisation) (1997) *Annual Report*. Cyprus Tourism Organisation, Nicosia.

CTO (Cyprus Tourism Organisation) (1999) *Annual Report*. Cyprus Tourism Organisation, Nicosia.

Dana, L.P. and Dana, T.E. (2000) Taking sides on the island of Cyprus. *Journal of Small Business Management* 38(2), 80–87.

Dodd, C.H. (ed.) (1993) *The Political, Social and Economic Development of Northern Cyprus*. Eothen Press, Huntingdon, UK.

Elliot, J. (1997) *Tourism*. Routledge, London.

Engelbert, P. (2000) Pre-colonial institutions, post-colonial states, and economic development in tropical Africa. *Political Research Quarterly* 53(1), 7–36.

Frendo, H. (1993) The legacy of colonialism: the experience of Malta and Cyprus. In: Lockhart, D.G., Drakakis-Smith, D. and Schembri, J. (eds) *The Development Process in Small Island States*. Routledge, London, pp. 151–160.

Godfrey, K.B. (1996) Towards sustainability? Tourism in the Republic of Cyprus. In: Harrison, L.C. and Husband, W. (eds) *Practising Responsible Tourism*. John Wiley & Sons, New York, pp. 58–80.

Gössling, S. (2000) Sustainable tourism development in developing countries: some aspects of energy use. *Journal of Sustainable Tourism* 8(5), 410–425.

Gunn, C.A. (1994) *Tourism Planning*. Taylor and Francis, Washington, DC.

Hall, D. (2000) Cyprus. In: Hall, D. and Danta, D. (eds) *Europe Goes East: EU Enlargement Diversity and Uncertainty*. The Stationery Office, London, pp. 155–167.

Hira, A. and Hira, R. (2000) The new institutionalism: contradictory notions of change. *The American Journal of Economics and Sociology* 59(2), 267–282.

Inskeep, E. (1991) *Tourism Planning*. John Wiley & Sons, New York.

Ioannides, D. (1992) Tourism development agents: the Cypriot resort cycle. *Annals of Tourism Research* 19(4), 711–731.

Ioannides, D. (2000) The dynamics and effects of tourism evolution in Cyprus. In: Apostolopoulos, Y., Loukissas, P. and Leontidou, L. (eds) *Mediterranean Tourism*. Routledge, London, pp. 112–128.

Ioannides, D. and Apostolopoulos, Y. (1999) Political instability, war, and tourism in Cyprus: effects, management, and prospects for recovery. *Journal of Travel Research* 38, 51–56.

Ivanko, J. (2001) Putting the 'eco' in tourism. *The Environmental Magazine* 12(1), 34–39.

Kibris Daily (2001) *Gecekondu Gibi,* 18 October, p. 10.

Lawrence, T.B. (1999) Institutional strategy. *Journal of Management* 25(2), 161–187.

Lockhart, D.G. (1993) Tourism and politics: the example of Cyprus. In: Lockhart, D.G., Drakakis-Smith, D. and Schembri, J. (eds) *The Development Process in Small Island States*. Routledge, London, pp. 228–246.

Lockhart, D.G. (1994) Tourism in Northern Cyprus: patterns, policies, and prospects. *Tourism Management* 15(5), 370–400.

Lockhart, D.G. (1997) Tourism to Malta and Cyprus. In: Lockhart, D.G. and Drakakis-Smith, D. (eds) *Island Tourism*. Pinter, London, pp. 152–180.

Mansfeld, Y. and Kliot, N. (1996) The tourism industry in the partitioned island of Cyprus. In: Pizam, A. and Mansfeld, Y. (eds) *Tourism, Crime and International Security Issues*. John Wiley & Sons, London, pp. 175–186.

MSDPM (Ministry of State and Deputy Prime Ministry) (1999) *Statistical Yearbook of Tourism*. Tourism Planning Office, Nicosia.

MTE (Ministry of Tourism and Environment) (2001) *Statistical Yearbook of Tourism*. Tourism Planning Office, Nicosia.

Nedjatigil, M.Z. (1982) *The Cyprus Conflict*. Tezel Offset and Printing Co. Ltd, Nicosia.

Nedjatigil, M.Z. (1996) The Turkish Republic of Northern Cyprus: statehood and recognition. *Journal for Cypriot Studies* 2(1), 3–19.

North, D. (1990) *Institutions, Institutional Change, and Economic Performance*. Cambridge University Press, New York.

North, D. (1995) The new institutional economics and Third World development. In: Harriss, J., Hunter, J. and Lewis, C. (eds) *The New Institutional Economics and Third World Development*. Routledge, London, pp. 17–26.

Olgun, M.A. (1993) Economic overview. In: Dodd, C.H. (ed.) *The Political Social and Economic Development of Northern Cyprus*. Eothen Press, Huntingdon, UK, pp. 270–298.

Panayiotopoulos, P.I. (1995) Cyprus: the developmental state in crisis. *Capital and Class* 57, 13–53.

Roberts, L. and Hall, D. (2001) *Rural Tourism and Recreation: Principles to Practice*. CAB International, Wallingford, UK.

Rustem, K. (1987) *North Cyprus Almanac.* Contact Type Setting Systems Ltd, London.

Saveriades, A. (2000) Establishing the social tourism carrying capacity for the tourist resorts of the east coast of the Republic of Cyprus. *Tourism Management* 21(2), 147–156.

Schlegel, J.F. (1999) Linking performance assessment to the strategic plan. *Association Management* 51(1), 75–79.

Scott, J. (1997) Chances and choices: women and tourism in Northern Cyprus. In: Sinclair, M.T. (ed.) *Gender, Work and Tourism.* Routledge, London, pp. 60–90.

Sharpley, R. (2003) Tourism, modernization and development on the island of Cyprus: challenges and policy responses. *Journal of Sustainable Tourism* 11(2/3), 246–265.

Sharpley, R. and Sharpley, J. (1997) *Rural Tourism.* International Thomson Business Press, London.

SPO (State Planning Organization) (1988) *Second Five-Year Development Plan (1988–1992).* State Planning Organization, Nicosia.

SPO (State Planning Organization) (1992) *Third Five-Year Development Plan (1993–1998).* State Planning Organization, Nicosia.

SPO (State Planning Organization) (1999) *Economic Developments in the Turkish Republic of Northern Cyprus 1998.* State Planning Organization, Nicosia.

Sreekumar, T.T. and Parayil, G. (2002) Contentions and contradictions of tourism as development option: the case of Kerala, India. *Third World Quarterly* 23(3), 529–548.

Theuns, H.L (2002) Tourism and development: economic dimensions. *Tourism Recreation Research* 27(1), 69–81.

Tosun, C. (2001) Challenges of sustainable tourism development in the developing world: the case of Turkey. *Tourism Management* 22(3), 289–303.

Vandenberg, P. (2002) North's institutionalism and the prospect of combining theoretical approaches. *Cambridge Journal of Economics* 26(2), 217–235.

Wilkinson, P.F. (1997) *Tourism Policy and Planning.* Cognizant Communication Corporation, New York.

Williams, D. (1998) Economic development and the limits of institutionalism. *SAIS Review* 18, 1–7.

World Bank (1992) *Governance and Development.* World Bank, Washington, DC.

World Bank (1997) *World Development Report.* World Bank, Washington, DC.

WTO (World Tourism Organization) (1994) *National and Regional Tourism Planning.* Routledge, London.

Yorucu, V. and Jackson, P.M. (1996) *Tourism and Environmental Planning in Small Island States: the Case of Northern Cyprus and Isle of Man.* University of Leicester, Leicester, UK.

11   The impact of political instability on the tourism industries of selected Mediterranean destinations: a neural network approach

Antonis Theocharous

Introduction

The impact of the terrorist attacks of 11 September 2001 in the USA placed the tourist industry, at an international level, in turmoil. They emphasized how important travel and tourism are to the global economy, but also how over-dependence on tourism can be economically and socially devastating. More than ever, there is a primary need for addressing issues related to sustainable tourism growth and crisis recovery.

Before 11 September, travel and tourism was the world's fastest-growing industry, accounting for one in every 12 jobs. When the massive $3.6 billion industry almost ground to a halt after the terrorist attacks, the ripple effects extended well beyond the USA, exposing the vulnerability of countries too dependent on international tourism. Total international arrivals in September 2001 dropped 29% when compared with the September 2000 figure (US Department of Commerce, 2001) and at the same time each of the top ten international markets experienced between 19 and 49% reductions (Table 11.1). International arrivals in September 2001 deepened the year-to-date decline to 6% below the same 9-month period in the previous year.

Despite considerable fluctuations in the growth rate over the years, the tourism indus-

Table 11.1. US international arrivals after the 11 September attack.

Country	September % change (2001/2000)	Jan–Sep % change (2001/2000)
Canada	−19	−4
Mexico	−30	3
Japan	−45	−7
United Kingdom	−27	−7
Germany	−46	−21
France	−34	−12
South Korea	−20	1
Brazil	−49	−12
Venezuela	−21	5
Italy	−40	−10

Sources: US Department of Commerce, US International Trade Administration.

try has demonstrated an extraordinary resistance and ability to overcome crises and disasters. Unquestionably, the 11 September attacks have had a more dramatic impact than any other crisis in recent years such as the Gulf War in 1991 and the terrorist attacks in Luxor, Egypt, in 1997.

In the last 50 years, tourism has been adversely affected by a wide range of problems – natural disasters, serious social conflicts, wars, economic crises and terrorism. The effects of 11 September have underlined that the symbiotic relationship between political instability and tourism needs to be understood and acted on, not just in terms of the fluctuation in tourism arrivals and receipts but also within a broader framework.

In light of the above, the aim of this chapter is to investigate the impact of political instability on the tourist industries of four eastern Mediterranean destinations – (southern) Cyprus, Greece, Turkey and Israel – during the period 1977–1997. It does this by producing a theoretical and empirical framework that will enable an evaluation of the relationship between political instability and tourism. These four countries are singled out because their tourist products are somewhat similar – sunny and warm climate, good beaches, but also with much cultural heritage – and they are situated at some distance from the major tourist-generating countries and, therefore, are better suited for longer holidays than for short weekend breaks.

Neural Network Simulations as an Impact Assessment Tool

In the last few years artificial neural networks have seen an explosion of interest and have been applied successfully to a wide range of disciplines such as medicine, engineering, finance and business (Verkooijen, 1996; Wong and Selvi, 1998; Krycha and Wagner, 1999; Vellido et al., 1999; Smith and Gupta, 2000; Wong et al., 2000).

Although neural networks originated in mathematical neurobiology as a description of brain function, their statistical basis renders them a successful modelling paradigm (Bishop, 1995). The popularity and success of the applicability of neural networks for analytical purposes can be mainly attributed to the following key characteristics:

- *Analytical power:* the fact that neural networks are sophisticated non-linear modelling techniques makes them capable of modelling extremely complex research functions.
- *Learning:* neural networks can be applied in cases where the form of the function is not known. Critical stages are the preparation of

data, selection of appropriate neural network model and algorithms, and interpretation of the results.
- *Generalization and noise tolerance:* a neural network is able to generalize in processing novel data, as well as adequately processing noisy input data that includes some level of error.

A neural network is composed of a large number of highly interconnected processing elements (neurones) working in parallel to solve specific problems. Thus instead of conceptualizing a problem as a purely mathematical one, neural networks use in combination a brain-like structure, information processing techniques and statistical optimization in order to develop a processing strategy (Hair et al., 1998).

Neural networks have the remarkable ability to derive meaning from complicated or imprecise data (Stergiou and Siganos, 1996). For this reason, the approach is used to extract patterns and detect trends that are too complex to be noticed by either humans or other computing techniques. In this context, a trained neural network can be thought of as an 'expert' in the category of information it has been given to analyse, and can be used to provide projections for new situations of interest and to answer 'what … if …?' questions.

The basic element of a neural network is the node (unit). The node, which is analogous to the neurone of the human brain, is a self-contained processing unit that acts in parallel with other nodes in the neural network by accepting inputs and computing outputs.

An important feature of a neural network is its ability to 'learn' or to 'correct itself' based on its errors. In using a biological analogy, the weights represent a state of a long-term memory – the model's 'best guess' as to how to make an interpretation of any input information. Once the input for a case is processed through the system, the resulted output can be compared to the required output value. Based on the difference between the two values, the model is adjusted within an iteration of improving its performance, so that on the occasion of the next input being presented, it will result in a smaller difference at the output. This is termed 'training the system in a supervised learning mode'. The objective is to process a large number of training examples through the neural network so that it can make the best functional mapping across all of the input data patterns.

Application of Neural Networks in Tourism Research

The application of neural networks in the field of tourism is still largely at an embryonic stage: a literature search led to the identification of just 11 articles. From these, only seven were relevant to tourism demand and forecasting (e.g. Pattie and Snyder, 1996; Law, 1999, 2000; Law and Au, 1999).

One of the most informative and important articles for the purposes of the current analysis is the study by Uysal and El Roubi (1999). The objectives of their study were to explore and demonstrate the usefulness of neural networks as an alternative approach to the use of multiple regression in tourism demand studies. Their results suggested that the application of neural networks in tourism demand studies is useful in identifying existing patterns that may not be revealed by multiple regression, and in some cases may result in better estimates in terms of prediction bias and accuracy.

The importance of Uysal's and El Roubi's research findings for the purposes of this study lies in the fact that they identified, either through their own work or by reviewing the relevant literature, the limitations of multiple regression that can be overcome by neural networks. Citing the studies that attempted to capture the relationship between political instability and tourism utilizing regression techniques (Rummel, 1963; Bwy, 1972; Sanders, 1981; Gupta, 1990; Enders and Sandler, 1998), Uysal and El Roubi (1999) provided strong evidence about the superiority of neural network models in capturing prediction and recognition of discriminating patterns. More specifically, the limitations of multiple regression that can be overcome by neural networks were defined as:

- the complexity of multiple regression models is limited to the linear combination of decision variables;

- the neural network models are not subject to model mis-specification as in the case with multiple regression, especially in short-term modelling;
- neural networks can partially transform the input data automatically, whereas this represents a rather exhaustive task with multiple regression; and
- neural network models are capable with their non-linear threshold functions of handling almost any kind of non-linearity, while this is not a quality performance guarantee with multiple regression models.

A Neural Network Model for the Evaluation of the Impact of Political Instability on Tourism

In this section a neural network model is developed to capture the impact of political instability on tourism. The justification for selecting demand measures (dependent variables) and determinants (independent variables) is well established and confirmed in the tourism demand literature (Crouch, 1992, 1994; Witt and Witt, 1995; Lim, 1997). This chapter, however, goes beyond the existing literature by attempting to integrate the dimensions of political instability within tourism demand models in order to investigate empirically the cause-and-effect relationships between political instability and tourism in the context of selected Mediterranean destinations.

A study by Seddighi et al. (2001) has empirically demonstrated that cultural background plays a significant role in the determination of the perceptions of tourists regarding the level of instability at selected Mediterranean destinations. The implication of this finding is that attitudes/perceptions of prospective holidaymakers from European destinations must not be treated as a unified sample. Thus for each tourism-generating country a dimensional structure should be extracted, instead of a single dimensional structure from a unified survey response sample.

A subsequent study by Theocharous (2002) has developed dimensional structures from two sets of attitudinal/perceptual data that were collected from the two leading European tourism-generating countries: the UK and Germany. In addition, a dimensional structure was extracted from the analysis of event data provided by POLINST. These three different groupings of political instability indicators are directly used for the purposes of this study.

Data collection and transformations

Secondary as well as primary sources of data were used. The selection of data for the neural network model was based on data availability, the reliability of data sources and measurability of the variables. The data on tourism arrivals for the countries under investigation were derived from the national tourism organizations (NTOs) of the selected countries. While all four countries employed can be considered as representative Mediterranean destinations, they provide a level of diversity in terms of political instability phenomena and political and social behaviour. Such diversity will allow examination of the neural network model under different conditions and the observation of different phenomena.

The data on political instability incidents are taken from coded data sets provided by POLINST database (Table 11.2) (Theocharous, 2002). POLINST is a stand-alone relational database detailing political instability events that took place in eight countries in the Middle East–Mediterranean region between 1977 and 1997. The event data were collected from numerous sources and were classified by taking into consideration the methods and goals for investigating the relationship between tourism and political instability. On this basis an index of political instability indicators was developed incorporating 28 indicators. POLINST database contains more than 7200 daily chronological entries of political instability events. Each event is characterized by coded variables such as date, type of action, number of fatalities, number of injuries, geographical location, organizations involved and bibliography of the sources from where the data were extracted. An embedded program has the ability to print out reports

and graphs and also offers background data on economic indicators of the countries, profiles of terrorist organizations and tourist arrivals and receipts. This database represents the most up-to-date and comprehensive data set available on political instability and tourism. Additionally, the fact that the database is based on a conceptual framework, which incorporates the concepts of political instability and tourism, rendered it to be the most appropriate to be used for the examination of the causal relationships between the two concepts.

To enable the use of POLINST data for analytical purposes, the data were aggregated to monthly totals providing 252 months of observation. As Bar-On (1996) noted, monthly data provide a more holistic picture of changes in the tourism environment in comparison with annual data. All events from 1977 to 1997, grouped by month of occurrence for each country under examination, were used for the analysis.

Figure 11.1 presents the number of events per month during this 252-month period.

Examination of the changes in the level of political instability over time yields some preliminary insights into the dynamics of the political environments of the countries under examination. First, for the entire 21-year period, the number of events presents an erratic pattern of fluctuations. Second, the data appear to be heavily distorted, leading to a preliminary conclusion that there are structural breaks in the data indicating a high degree of non-linearity. This conclusion is of considerable importance since economic theory suggests that time series with structural breaks (non-linear) are considered to be inappropriate for traditional econometric time series analysis; use of neural networks can provide better and more accurate results.

The data for the remaining variables (income, exchange rates, price of oil, consumer price index) were collected from the following sources.

● Dataset No. 2618 Balance of Payments Statistics, International Monetary Fund 1965–Present. The dataset was accessed through an agreement provided by MIMAS[1] – University of Manchester and Data Archive, University of Essex. The datasets

[1] MIMAS (Manchester Information & Associated Services) is a JISC-supported national data centre providing the UK higher education, further education and research community with networked access to key data and information resources to support teaching, learning and research across a wide range of disciplines.

Table 11.2. Factors of the POLINST model.

F1POL	Armed attack events
	Terrorist attacks or threats of attacks to non-tourist targets (foiled bombing, sabotage)
	Assassinations (political)
	Kidnappings
	Mass arrests
	Political illegal executions
	Riots
	Imposition of martial law
	Arrests of significant persons
	Guerrilla warfare
F2POL	Peaceful demonstrations
	Restriction of political rights
	Peaceful strikes
	Censoring of media
F3POL	Hijacking
	Threat of war with another country
	Bombings
F4POL	War
	Unsuccessful *coups d'état*
	Army attacks beyond the country's borders
F5POL	Change in government
	Change in the political party governing a country
F6POL	Political instability in neighbouring countries
	War in a neighbouring country
F7POL	Terrorist attacks to tourism industry-related targets (sightseeing, transportation, accommodation)
	Successful *coups d'état*
	Terrorist attacks involving tourists as victims

were network accessed and X-Windows interface was used to employ the datasets and extract the data from SAS format.

- Dataset No. 2678 Government Finance Statistics, 1971–1997; International Monetary Fund. Network accessed from MIMAS.
- Dataset No. 2731 International Financial Statistics, 1948–July 1991; International Monetary Fund. Network accessed from MIMAS.
- OECD Main Economic Indicators Dataset 1960–present. Network accessed from MIMAS.
- Monthly Energy Market Chronology 1970–present. United States Department of Energy.

- State of Israel – Central Bureau of Statistics.
- Republic of Cyprus – Statistical Service.

For every country under examination (Cyprus, Greece, Israel and Turkey) the demand for travel was expressed in three separate models as follows:

$$Y = f\ (INC, ER, PO, CPI, F1GER, F2GER, F3GER, F4GER)$$
$$Y = f\ (INC, ER, PO, CPI, F1UK, F2UK, F3UK, F4UK)$$
$$Y = f\ (INC, ER, PO, CPI, F1POL, F2POL, F3POL, F4POL, F5POL, F6POL, F7POL)$$

where Y = the number of tourist arrivals in destination X; INC = average *per capita* income of tourists for five major European tourism-generating countries; ER = the foreign exchange rate (national currency/US$); PO = price of oil; CPI = Consumer Price Index at the destination as a proxy to the cost of living at the destination; $F1GER–F4GER$ denote the factors of political instability that resulted from the analysis of the German survey sample; $F1UK–F4UK$ denote the factors of political instability that resulted from the analysis of the UK survey sample; and $F1POL–F7POL$ denote the factors of political instability that resulted from the analysis of POLINST dataset.

The justification for selecting demand measures (dependent variables) and determinants (independent variables) is well established and confirmed in the literature on tourism demand studies (Johnson and Ashworth, 1990; Crouch, 1992, 1994; Witt and Witt, 1992; Lim, 1997; Song and Witt, 2000). The review of these studies has suggested that the exogenous variables for international tourism demand comprise mainly: the income of the origin country, the cost of living in the destination country, currency foreign exchange, marketing expenditure on promotional activities in the destination country and transportation costs.

The CPI has been used frequently as a reasonable proxy for the cost of tourism (Crouch, 1992; Lim, 1997). The use of CPI is justified on the grounds of convenience (the data are widely available) and the argument that tourist spending is spread over a wide range of the economy and so may approximate the general average consumer spending

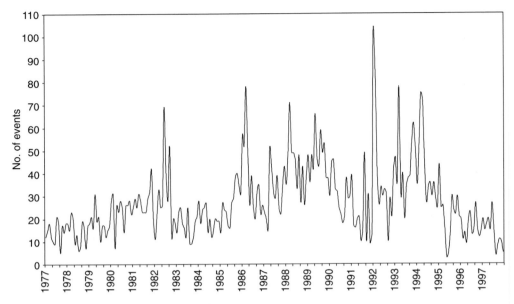

Fig. 11.1. Total number of events per month, 1977–1997.

patterns (Uysal, 1998; Uysal and El Roubi, 1999). In addition, Crouch (1992) argued that the mix of goods and services consumed by tourists is not too different from the mix constituting the CPI, and that the changes in the CPI fairly reflect changes in the prices of goods and services consumed by tourists. It is therefore safe to assume that the CPI will track tourism and travel prices closely.

It should be noted that the CPI for two of the countries (Greece and Turkey) was obtained from the OECD datasets. The base year for the data reported from 1977 to 2000 was 1995, therefore no transformations were needed. The same stands for Israel where the base year for the CPI reported by the Israeli Central Bureau of Statistics from 1951 to 2001 was 1951. The only transformation that had to be made was for Cyprus, where for the data reported from 1977 to 1992 the base year was 1985, and 1992 was the base year for data reported from 1992 onwards. To solve this problem the indices were chained with data for a base being used from that year until the year before the next base year. Therefore, the chaining coefficient between the 2 successive base years (1985 and 1992) is equal to the value of

the annual index (1992), with base 1985=100 divided by 100. By following this approach the chaining coefficient for Cyprus was calculated (1.305) and the data were subsequently adjusted.

The price of oil variable was used as a proxy for transport costs from the tourism-generating country to the country of destination. Furthermore, the values for the variable were converted to reflect percentage annual change by using the following formula:

$$\% \text{ Change at time } t = \text{Change } Y/Y_{t-1} = (Y_t - Y_{t-1})/Y_{t-1} = (Y_t/Y_{t-1}) - 1$$

The resulting figures were then multiplied by 100 to scale up to a whole number.

The exchange rate as an explanatory variable in the demand function was justified in that consumers are more aware of exchange rates than prices in the destination (Witt and Witt, 1995). This suggests that the price of the currency in the destination country will influence tourists' expenditures, *ceteris paribus*. Thus, if the price of the foreign currency declines, tourists are likely to demand more services and/or spend more on services.

The data on income were derived by calculating the average income *per capita* of the six major European tourism-generating countries (United Kingdom, Germany, France, Italy, the Netherlands and Switzerland). Because the *per capita* income figures were reported in national currencies, they were converted to US$ by using the corresponding exchange rate parity for each month, and then were averaged so as to obtain a single monthly series for all the tourism-generating countries. This was considered as the most appropriate solution since the use of country-specific series would demand the development of several country-specific tourism demand models.

Pre-processing of data

As a crucial part of the neural network modelling approach, the data had to be subjected to a pre-processing stage of normalization.

The dependent variable (tourist arrivals) as well as the first four explanatory variables (exchange rate, consumer price index, price of oil, income) were normalized with respect to the mean value of each variable. Prior to the normalization, the change in the value since the last year was computed. The following formula was used for the normalization of the tourist arrivals: $Y^{i,j}$ for month i ($i=1,...,12$) in year j ($j=1,...,21$):

$$Y^{i,j}_{norm} = \frac{Y^{i,j} - Y^{i,j-1}}{Y^{i}_{mean}}$$

where Y_{mean} is the mean (average) number of arrivals for month i ($i=1,...,12$). Analogous equations were used for the other four variables.

For the three sets of factors of political instability (UK, Germany, POLINST) (Tables 11.3, 11.4 and 11.2) the normalization procedure was based on the maximum value of each factor (variable) and reflected the absolute value of each variable for a particular month:

$$F^{i,j}_{norm} = \frac{F^{i,j}}{F^{i,j}_{max}}$$

where F_{max} is the maximum value of the political instability factor for month i ($i=1,...,12$).

Table 11.3. Factors of the UK model.

F1UK	Political illegal executions
	Unsuccessful *coups d'état*
	Kidnappings
	Hijackings
	War in a neighbouring country
	Terrorist attacks or threats of attacks to non-tourist targets (foiled bombing, sabotage)
	Political instability in neighbouring countries
	Assassinations (political)
	Armed attack events
	Army attacks beyond the country's borders
	War
	Successful *coups d'état*
	Bombings
	Threat of war with another country
F2UK	Peaceful strikes
	Arrests of significant persons
	Peaceful demonstrations
	Change in the political party governing a country
	Restriction of political rights
F3UK	Mass arrests
	Censoring of media
	Guerrilla warfare
	Riots
	Imposition of martial law
	Terrorist attacks to tourism industry-related targets (sightseeing, transportation, accommodation)
	Civil war
F4UK	Terrorist attacks involving tourists as victims
	Change in government

The normalized data were used for training and testing the neural network on its ability to predict the number of tourist arrivals. The above normalization of the dependent variable tackles implicitly the problem of seasonality in tourism. The model deals with the change in the number of tourist arrivals for a particular month compared to the same month in the previous year. Furthermore, any regular seasonal fluctuations are already embedded in the value for the previous month. Hence such seasonal fluctuations will be implicitly present in the final number of tourist arrivals for the current month. In contrast, if the change over the previous month is used, which is a common practice, any periodic fluctuations will have to be explicitly calculated by the model.

Table 11.4. Factors of the Germany model.

F1GER	Restriction of political rights
	Censoring of media
	Mass arrests
	Political illegal executions
	Arrests of significant persons
	Assassinations (political)
F2GER	Army attacks beyond the country's borders
	Terrorist attacks or threats of attacks to non-targets (foiled bombing, sabotage)
	War in a neighbouring country
	War
	Terrorist attacks to tourism industry-related targets (sightseeing, transportation, accommodation)
	Kidnappings
	Threat of war with another country
	Armed attack events
	Imposition of martial law
	Hijacking
	Bombings
F3GER	Change in government
	Peaceful demonstrations
	Change in the political party governing a country
	Peaceful strikes
	Political instability in neighbouring countries
F4GER	Unsuccessful *coups d'état*
	Riots
	Civil war
	Successful *coups d'état*
	Terrorist attacks involving tourists as victims
	Guerrilla warfare

The neural network architecture and training

There are two main approaches in processing time series with neural networks. The sequential processing approach requires recurrent structures in the network architecture, such as recurrent layers and connection, which will be able to carry information between the elements of the serial pattern (Panchev and Wermter, 2000; Wermter and Panchev, 2000). Another approach, which was applied in this study, is the so-called sliding time window (Bishop, 1995) which relies on the assumption that the time series depend on explanatory variables and previous values of the dependent variable from a finite number of time steps in the past. The assumption taken in the current model

was that the number of tourist arrivals depends primarily on the values of the explanatory factors from the last 12 months. Bearing in mind the seasonal character of tourism, such an assumption is justified.

The neural network used for the experimental setup with UK factors is presented in Fig. 11.2. The neurones in the input layer are grouped in 12 different sets feeding in information about the 12 months previous to the current one. Each set receives information for the change in the number of tourist arrivals, for the explanatory variables as well as the political instability factors.

After processing a given input, the output of the network contains a single value representing the predicted change of the tourist arrivals for the current month. A semantic interpretation of the network processing in the context of the prediction task would be: what is the change in the number of tourist arrivals, given its change for each of the last 12 months, the change in the explanatory economic indicators for each of the last 12 months and the relative number of political instability events that happened in each of the last 12 months?

The available normalized data of 19 years (228 observations) were split into a test set containing data for the years 1980, 1990, 1995, 1996 and 1997, and a training set containing the data for the rest of the years. Such a choice of training and test data allows for the examination of the model for the tourist arrivals at the beginning, the middle and the end of the observed period. Structural breaks in the data were also present in both training and test data. In other words, the performance of the model could be examined throughout the period of available data, as well as in extreme conditions of structural breaks.

Results

Figures 11.3–11.5 present graphically the results of the neural network model for the factors of political instability as resulted from the analysis of the German sample for Cyprus, Israel and Turkey.

An examination of the figures shows that the neural network model in all three cases follows quite satisfactorily the pattern of actual

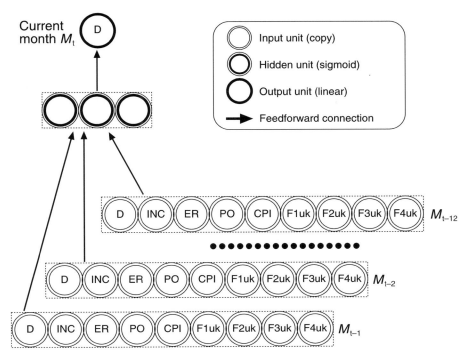

Fig. 11.2. Neural network architecture.

arrivals. It is also worth noting that the prediction is less accurate when the factors for political instability are not incorporated in the model. This becomes evident when the network is tested without the four political instability factors (grey line). A preliminary conclusion can be then extracted suggesting that the factors contribute towards the overall predicting efficiency of the model.

By comparing the three countries it can be ascertained that Turkey has the most accurate results, as indicated by the accuracy measurements of the three models compared. This accuracy measurement is based on mean absolute percentage error (MAPE) and normalized correlation coefficient (r).

To calculate MAPE the following formula was used:

$$\text{MAPE} = \frac{\sum_{i=1}^{n} \frac{X_i - Y_i}{Y_i}}{n} * 100\%$$

where X_i and Y_i represent the estimated and actual tourist arrivals for $i=1,\ldots,228$.

The normalized correlation coefficient r measures the closeness of the observed and estimated arrivals.

The formula that was used to calculate r is as follows:

$$r = \frac{\sum_{i=1}^{n}\left(X_i * Y_i\right)}{\sqrt{\sum_{i=1}^{n}\left(X_i\right)^2 * \sum_{i=1}^{n}\left(Y_i\right)^2}}$$

where X_i and Y_i represent the estimated and actual tourist arrivals for $i=1,\ldots,228$.

By contrasting the MAPE figures for all three countries, it can be concluded that Turkey has the lowest MAPE (14.45%) while Cyprus and Israel have 16.62% and 16.46% respectively. The low MAPE indicates that the deviations between the discrepancies among the predicted values, derived by the neural network, and the actual values are very small. According to Law and Au (1999), for a MAPE within a 15% discrepancy range, the neural network succeeds in achieving approximately 80% of the output in the acceptable range. In addition, the normalized correlation of the coefficient between the predicted and actual

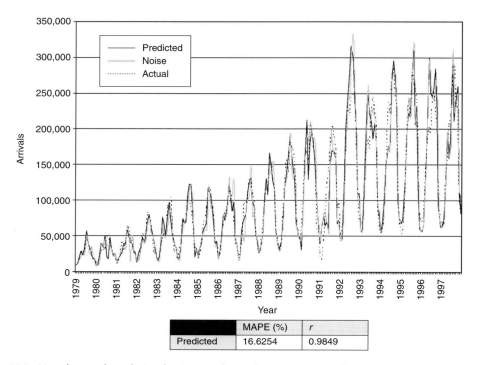

	MAPE (%)	r
Predicted	16.6254	0.9849

Fig. 11.3. Neural network prediction for Cyprus – factors from German sample.

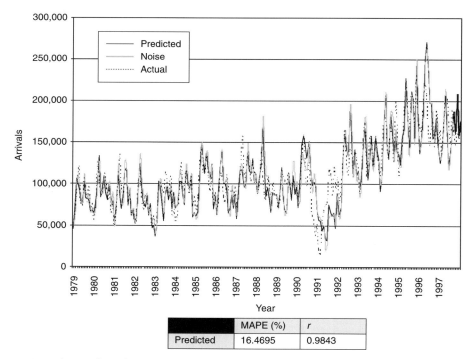

	MAPE (%)	r
Predicted	16.4695	0.9843

Fig. 11.4. Neural network prediction for Israel – factors from German sample.

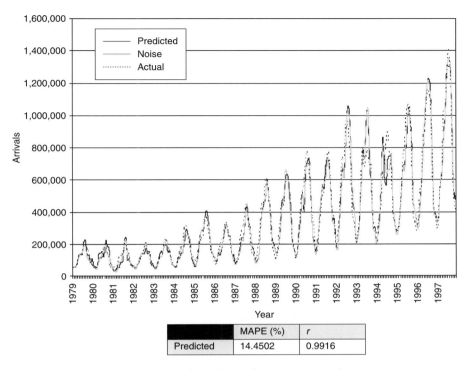

	MAPE (%)	r
Predicted	14.4502	0.9916

Fig. 11.5. Neural network prediction for Turkey – factors from German sample.

values (r) is very high for all three countries, with values of almost one. Again, as in the case with MAPE, Turkey appears to have the highest value of r when compared with Cyprus and Israel. A high value of r is an indicator of the close relationship between the estimated results and the actual tourism data.

The results of the neural network model for the factors of political instability, as provided from the analysis of the UK sample for Cyprus, Israel and Turkey, are presented in Figs 11.6–11.8.

As in the case of the factors that resulted from the analysis of the German sample, the neural network model appears to capture well the actual tourism arrivals trend. The behaviour of the model (predicted, as well as noise), as tested with the UK factors, resembles to a great extent the behaviour of the model with the German factors.

A notable deviation as compared with the previous model is the higher MAPE for Cyprus. However, the difference is minor and it does not undermine in any way the predictability of

the neural network model. Furthermore the reported r values are very high, reaffirming the close relationship between the predicted and actual data.

The last set of tests (Figs 11.9–11.12) was performed with neural networks trained with the factors that have resulted from the analysis of the POLINST dataset.

The performance of the model is significantly better in comparison with the previous two models (the German and the UK). The curves of the actual and predicted data appear to converge and follow the same pattern. These first findings lead to a preliminary conclusion that the training of a neural network, by using factors which result from actual data (in the form of political instability events) instead of those which result from perceptual data (in the form of opinions/perceptions/attitudes), is much more promising and in extent more reliable and valid.

By examining more carefully the figures, and in particular the noise curve, it can be ascertained that the overall predictive accuracy

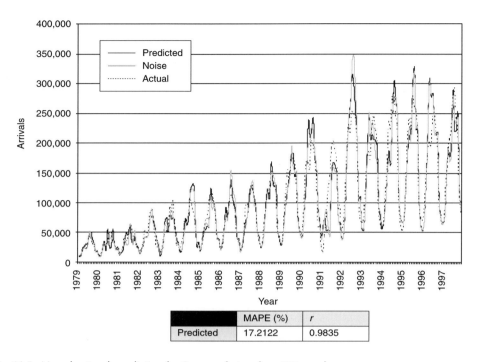

	MAPE (%)	r
Predicted	17.2122	0.9835

Fig. 11.6. Neural network prediction for Cyprus – factors from UK sample.

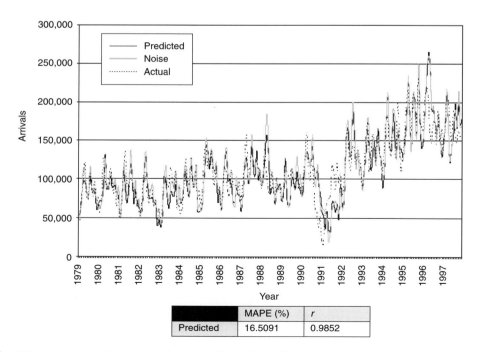

	MAPE (%)	r
Predicted	16.5091	0.9852

Fig. 11.7. Neural network prediction for Israel – factors from UK sample.

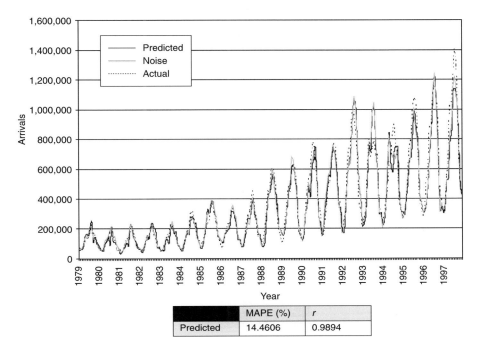

	MAPE (%)	r
Predicted	14.4606	0.9894

Fig. 11.8. Neural network prediction for Turkey – factors from UK sample.

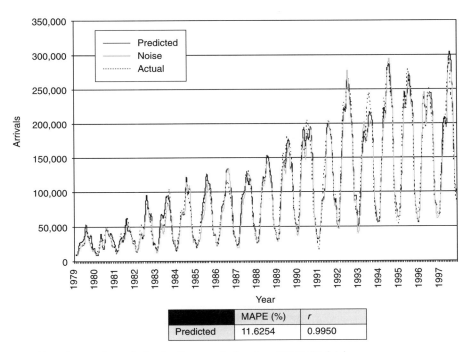

	MAPE (%)	r
Predicted	11.6254	0.9950

Fig. 11.9. Neural network prediction for Cyprus – factors from POLINST database.

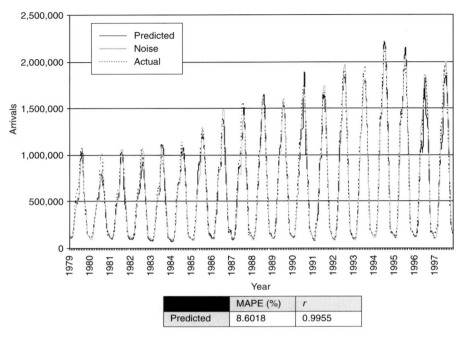

	MAPE (%)	r
Predicted	8.6018	0.9955

Fig. 11.10. Neural network prediction for Greece – factors from POLINST database.

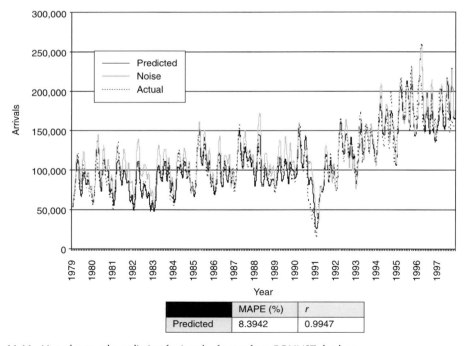

	MAPE (%)	r
Predicted	8.3942	0.9947

Fig. 11.11. Neural network prediction for Israel – factors from POLINST database.

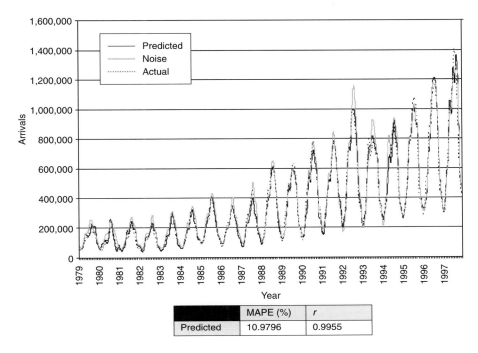

	MAPE (%)	r
Predicted	10.9796	0.9955

Fig. 11.12. Neural network prediction for Turkey – factors from POLINST database.

of the model is significantly better when the model is tested including the POLINST factors. This is supported by the fact that when the political instability factors are excluded (the grey curve), there is a clear deviation from the pattern that is followed by the dotted (actual) and black (predicted) curves. From the four country models the most representative of the above argument is Israel. This is not surprising at all, considering that 45% of all political instability events, which are reported in POLINST, have taken place in Israel. Therefore, considering the high frequency of such events in Israel, it can be suggested that any segregation of these events from the neural network will inevitably cause serious digression from the normal prediction (see Fig. 11.11). This result indicates the importance of the political instability factors for the performance of the prediction model.

The clear convergence of the dotted (actual) and black (predicted) curves reveals that the number of tourist arrivals estimated by the neural network is very close to the actual

values. In other words, the predicted output of the model appears to be accurate, with a relatively small chance of error.

By examining the MAPE values for all four countries it can be clearly seen that they are much lower than those of the German and UK factors. This indicates that for the POLINST neural network model the deviations between the discrepancies among the predicted values (derived from the neural network) and the actual values are very small. All four reported MAPE values are under 12% discrepancy rate with Greece and Israel having the lowest: 8.60% and 8.39% respectively. All reported values for the normalized correlation coefficient (r) are over 0.994, demonstrating the close relationship between the estimated and the actual tourist arrivals.

Importance of factors

The aim in this section is to test the neural network by omitting each of the input factors. The importance of a factor was evaluated as being proportional to the neural network's output error during tests, with the data of that factor being removed. The output error is calculated as:

$$\sum_{i=1}^{228} \left(y_i - d_i\right)^2$$

where y is the predicted number of arrivals and d is the actual (desired) number. The output error was calculated over the entire dataset. By comparing the error values of the different factors, it was possible to draw conclusions about the level of importance of a given factor for the performance of the model, and therefore for its influence on the predicted number of arrivals. When a particular input factor is removed, a higher output error will indicate whether this factor is relatively influential on a correct output of the model. Respectively, lower error means that the absence of the factor does not degrade the performance, and leads to the conclusion that it is less important.

Since the results from the above section demonstrated that the POLINST model is the one with the highest levels of predictive accuracy, the experiments based on the importance of each of the factors were based on that model.

The results presented in Figs 11.13–11.16

are clear indications for the significance of the POLINST factors for the overall performance of the model. It must be noted that the values, which are reported on the graphs, are mainly indicative and allow only comparative qualitative analysis but not quantitative conclusions. What is of importance here is the dominating influence of the POLINST factors on the overall predictive power of the model. Although the practice that was followed here is novel in the tourism domain, the findings are informative and provide a clear indication about the contribution of the independent variables towards the overall model prediction.

The removal of POLINST factors from the model of Israel (Fig. 11.13) produces the highest output error compared with the other countries. This clearly indicates that this set of factors is very influential on the performance of the model for Israel. POLINST factors for Turkey (Fig. 11.15) have similar influence. These findings are not surprising since both Israel and Turkey have both suffered from prolonged periods of internal conflict and instability, which inevitably have left their consequences on the tourism industry. For the remaining two countries, Cyprus and Greece, the POLINST factors (Figs 11.14 and 11.16) appear to be influential to a lesser extent.

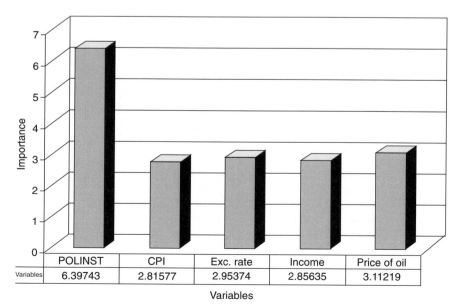

Variables	POLINST	CPI	Exc. rate	Income	Price of oil
Variables	6.39743	2.81577	2.95374	2.85635	3.11219

Variables

Fig. 11.13. Israel – importance of variables for model prediction.

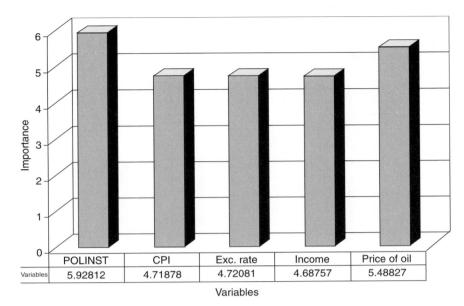

Variables	POLINST	CPI	Exc. rate	Income	Price of oil
	5.92812	4.71878	4.72081	4.68757	5.48827

Variables

Fig. 11.14. Cyprus – importance of variables for model prediction.

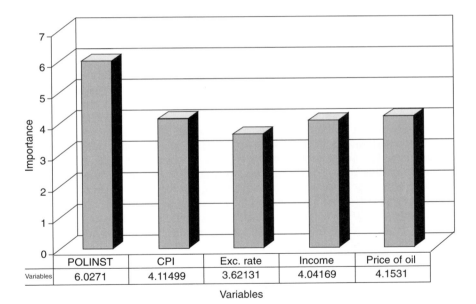

Variables	POLINST	CPI	Exc. rate	Income	Price of oil
	6.0271	4.11499	3.62131	4.04169	4.1531

Variables

Fig. 11.15. Turkey – importance of variables for model prediction.

Time lag for events of political instability

The most representative case of the influence of political instability from the previous section, i.e. Israel, was considered for the final experi-

ment, where the time lag for events of political instability was examined. Similarly to the previous set of experiments, in a sequence of tests, the data of political instability for each of the preceding 12 months were suppressed from

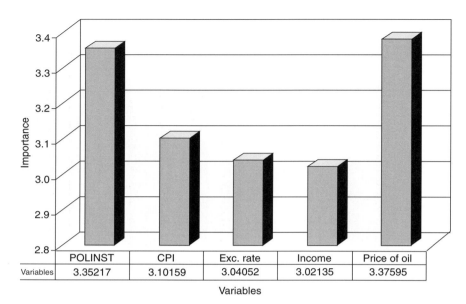

Variables	POLINST	CPI	Exc. rate	Income	Price of oil
	3.35217	3.10159	3.04052	3.02135	3.37595

Variables

Fig. 11.16. Greece – importance of variables for model prediction.

the input of the network. The resultant output error for each of the 12 months was recorded and compared (Fig. 11.17). As the results suggest, the neural network considers the most influential political instability events as those that have occurred in the previous 2 months.

It has to be noted that the results do not underestimate the impact of the other months on the prediction performance (in fact, they all generate relatively high output error), therefore it is not suggested that the other months are not important. What can be concluded is that the highest impact of any major political instability events is felt after approximately 2 months.

Although the methodology which is adopted here is novel, the results are compatible with those found by Enders and Sandler (1991) in their study of terrorism and tourism in Spain. By using VAR analysis they attempted to estimate the impact of terrorism on tourism in Spain during the period from 1970 to 1980. As in the current study, they have also used 12-month lags composed of monthly data. Their findings suggested that after a 'typical' terrorist incident, tourism to Spain began to decline at the beginning of the third month.

Conclusion and Policy Implications

The application of neural networks has demonstrated, clearly, that they have the analytical ability to provide accurate predictions and the flexibility to incorporate various forms and types of independent variables, that might be present in a tourism demand function.

The review of the relevant literature has demonstrated that the traditional econometric/statistical techniques do not have the ability to recognize the hidden patterns that are present in non-linear data, characterized with structural breaks and drastic changes. Based on the success of the presented model, it can be concluded that the incorporation of neural networks for the study of tourism demand, in the context of political instability, has provided a valid approach to tourism demand modelling and a spur to further research.

The results of the models, which were developed, confirm the current opinion that events of political instability can have a strong influence on the tourism industry. The series of experiments revealed further details of the inter-relationship between the political instability factors and the number of tourist arrivals.

To the best of our knowledge, the present

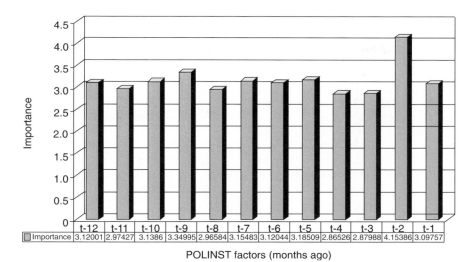

POLINST factors (months ago)

	t-12	t-11	t-10	t-9	t-8	t-7	t-6	t-5	t-4	t-3	t-2	t-1
Importance	3.12001	2.97427	3.1386	3.34995	2.96584	3.15483	3.12044	3.18509	2.86526	2.87988	4.15386	3.09757

Fig. 11.17. Time lag for events of political instability (predicting month M_t).

study is the first successful attempt to model the relationship of political instability to tourism in a neural networks analytical framework. The results provide confidence for further research and improvements. The importance of the current study is in its clear demonstration that the neural network model represents an effective approach towards modelling tourism demand and offers an important computational paradigm for tourism researchers who are interested in the prediction of cause and effect relationships.

Undoubtedly, for an issue such as political instability, the close monitoring, assessment and evaluation of its possible impacts are vital for tourism policy makers in order to develop and/or readjust their business policies. The neural network model which has been developed in this study, will, from a managerial point of view, provide valuable information regarding the impact of the various forms of political instability on tourism.

The sensitive profit margins of the tourism industry do not allow for mistakes that can be a result of wrong evaluations and assessments. Thus, sophisticated modelling such as neural networks can be of valuable assistance in providing reliable and valid long-term strategic planning.

The ever-changing global and regional socio-political dynamics that the tourism indus-

try faces have demonstrated the extent of its susceptibility to exogenous factors such as political and economic instability. This is particularly evident in the aftermath of the 11 September atrocity and the war in Iraq.

The countries that were examined are trying hard to devise and develop policies and strategies to counterbalance the negative effects that these events have brought, and at the same time reverse the negative climate. Tourism professionals and policy makers have at last recognized the need, significance and importance of systems for predicting and assessing the risks to which the tourism industry is susceptible. This imperative need for a framework for the continuous monitoring and assessment of socio-political dynamics in the tourism industry is now being responded to at the highest level both in the public and private sectors.

At the same time this research, and in particular the POLINST database, has provided the spark for the development of a sophisticated risk assessment system for the ongoing monitoring, assessment and forecasting of political instability at tourism destinations worldwide. This project is expected to provide an extensive, accurate and objective database of events of political instability that would allow the quantification of the relationship between political instability and tourism on an ongoing basis. Through the development of a neural network Internet agent, the project will allow the machine coding of events of political instability, and at a next step to use the data set so as to evaluate destinations worldwide in relation to their relative safety. This will enable the provision of ongoing assessments and forecasts regarding the relative levels of stability/instability at tourist destinations.

In this way, the significance of this work transcends issues of 'transition' and reflects the wider need for accessible and robust methodological and analytical tools to respond to the ever-changing global conditions that tourism development faces. None the less, the focus of this chapter on the variously 'transitional' societies of the eastern Mediterranean has been particularly apposite for this volume.

Acknowledgements

The author wishes to thank Derek Hall and Vivian Kinnaird for their helpful comments and suggestions on earlier versions of this study.

References

Bar-On, R. (1996) Measuring the effects on tourism of violence and of promotion following violent acts. In: Pizam, A. and Mansfeld, Y. (eds) *Tourism, Crime, and International Security Issues*. John Wiley & Sons, Chichester, UK, pp. 159–174.

Bishop, C. (1995) *Neural Networks for Pattern Recognition*. Clarendon Press, Oxford.

Bwy, D.P. (1972) Political instability in Latin America: the cross-cultural test of a causal model. In: Feierabend, I.K. and Gurr, T.R. (eds) *Anger, Violence and Politics*. Prentice-Hall, Englewood Cliffs, New Jersey.

Crouch, G. (1992) Effect of income and price on international tourism. *Annals of Tourism Research* 19(4), 643–664.

Crouch, G. (1994) The study of international tourism demand: a survey of practice. *Journal of Travel Research* 33(1), 12–31.

Enders, W. and Sandler, T. (1991) Causality between transnational terrorism and tourism: the case of Spain. *Terrorism* 14, 49–58.

Enders, W. and Sandler, T. (1998) *Transnational Terrorism in the Post-Cold War Era*. Department of Economics, Iowa State University, Iowa.

Gupta, D. (1990) *The Economics of Political Instability: The Effect of Political Instability on Economic Growth*. Praeger, New York.

Hair, J., Anderson, R., Tatham, R. and Black, W. (1998) *Multivariate Data Analysis*, 5th edn. Prentice Hall, Upper Saddle River, New Jersey.

Johnson, P. and Ashworth, J. (1990) Modelling tourism demand: a summary review. *Leisure Studies* 9, 145–160.

Krycha, K.A. and Wagner, U. (1999) Applications of artificial neural networks in management science: a survey. *Journal of Retailing and Consumer Services* 6, 185–203.

Law, R. (1999) Room occupancy rate forecasting: a neural network approach. *International Journal of Contemporary Hospitality Management* 10(6), 234–239.

Law, R. (2000) Back-propagation learning in improving the accuracy of neural network-based tourism demand forecasting. *Tourism Management* 21, 331–340.

Law, R. and Au, N. (1999) A neural network model for forecasting Japanese demand for travel to Hong Kong. *Tourism Management* 20, 89–97.

Lim, C. (1997) Review of international tourism demand models. *Annals of Tourism Research* 24(2), 835–849.

Panchev, C. and Wermter, S. (2000) Complex preferences for the integration of neural codes. In: *Proceedings of the International Joint Conference on Neural Networks*, Como, pp. 253–258.

Pattie, D. and Snyder, J. (1996) Using a neural network to forecast visitor behaviour. *Annals of Tourism Research* 23(1), 151–164.

Rummel, R. (1963) Dimensions of conflict behaviour within and between nations. *General Systems Yearbook* 8, 1–50.

Sanders, D. (1981) *Patterns of Political Instability*. Macmillan Press, London.

Seddighi, H.R., Nuttall, M.W. and Theocharous, A.L. (2001) Does cultural background of tourists influence the destination choice? An empirical study with special reference to political instability. *Tourism Management* 22(2), 181–191.

Seddighi, H., Theocharous, A. and Nuttall, M. (2002) Political instability and tourism: an empirical study with special reference to the microstate of Cyprus. *International Journal of Hospitality Management* 3(1), 12–24.

Smith, K.A. and Gupta, J.N.D. (2000) Neural networks in business: techniques and applications for the operations researcher. *Computer & Operations Research* 27, 1023–1044.

Song, H. and Witt, S.F. (2000) *Tourism Demand Modelling and Forecasting. Modern Econometric Approaches*. Pergamon, New York.

Stergiou, C. and Siganos, D. (1996) Neural networks, surveys and presentations in information systems engineering. *SURPRISE* 4, 1–21.

Theocharous, A.L. (2002) A Theoretical and Empirical Framework for the Examination of the Relationship between Political Instability and Tourism with Special Reference to Selected Mediterranean Destinations. Unpublished PhD thesis, University of Sunderland, Sunderland, UK.

US Department of Commerce, International Trade Administration (2001) *TI News announcement – International arrivals drop 29 percent in September*. US Department of Commerce, International Trade Administration, Washington, DC.

Uysal, M. (1998) The determinants of tourism demand: a theoretical perspective. In: Ioannides, D. and Debbage, K. (eds) *The Economic Geography of Tourism*. Routledge, London, pp. 79–95.

Uysal, M. and Crompton, J.L. (1985) An overview of approaches used to forecast tourism demand. *Journal of Travel Research* 23(4), 7–15.

Uysal, M. and El Roubi, S. (1999) Artificial neural networks versus multiple regression in tourism demand analysis. *Journal of Travel Research* 38, 111–118.

Vellido, A., Lisboa, P.J.G. and Vaughan, J. (1999) Neural networks in business: a survey of applications (1992 – 1998). *Expert Systems with Applications* 17, 51–70.

Verkooijen, W.J.H. (1996) *Neural Networks in Economic Modelling*. Tilburg University Press, Tilburg.

Wermter, S. and Panchev, C. (2000) Hybrid sequential machines based on neuroscience. *Proceedings of the ECAI Workshop on Foundations of Connectionist Symbolic Integration, Berlin*, pp. 13–24.

Witt, S.F. and Witt, C.A. (1992) *Modelling and Forecasting Demand in Tourism*. Academic Press, London.

Witt, S.F. and Witt, C.A. (1995) Forecasting tourism demand: a review of empirical research. *International Journal of Forecasting* 11(3), 447–475.

Wong, B.K. and Selvi, Y. (1998) Neural network applications in finance: a review and analysis of literature (1990–1996). *Information and Management* 24, 129–139.

Wong, B.K., Lai, V.S. and Lam, J. (2000) A bibliography of neural network business applications research: 1994–1998. *Computer & Operations Research* 27, 1045–1076.

12 Provincial government roles in Chinese tourism development: the case of Hunan

Rong Huang

Introduction

As one of the provinces of central China, Hunan has a population of 64.3 million (more than most European countries or US states), consisting of 42 nationalities and a territory of 210,000 sq km. With a recorded history of over 3000 years, it is well endowed with cultural and natural tourist resources.

Tourism in Hunan is still a relatively new phenomenon, however: it had not assumed any significance in the government's agenda nor had it been developed into a mass industry before 1978 (Qu, 1999). Since the introduction of the economic reform ('open door') policy by Deng Xiao-ping in 1978, tourism in Hunan has developed rapidly and has become a significant industry. In 2001, visitor arrivals totalled 51.5 million: 500,000 inbound tourists and 51 million domestic. Tourism receipts reached 21 billion RMB, equivalent to US$2.625 billion (CNTA, 2003).

Government involvement has greatly influenced tourism development in many countries, particularly those less developed countries with socialist political systems. Governments in developing countries tend to be more actively involved and assume key developmental and operational roles. In socialist countries, where the private sector may be small or non-existent, the level of government involvement is greater than in countries that have a predominantly free-enterprise philosophy (D. Hall, 1991; C.M. Hall, 1994).

China is a developing socialist country, and the private sector in the tourism industry is small. The rapid growth of tourism in Hunan since 1978 is largely due to the efforts of the Hunan provincial government: most aspects of tourism development have been addressed through a variety of policy initiatives and measures.

Although the government has been decisive in guiding Hunan's tourism development, there is little understanding about the types of roles played by the provincial government in tourism research literature. Without identifying the nature of the government's role, neither can past tourism development in Hunan be systematically analysed, nor can future trends be forecast.

The purpose of this chapter is first to identify the different roles that the Hunan government has played in the development of tourism during three recognized periods (Zhang, 1995) since 1978. Secondly, the author forecasts the different roles that the Hunan government is likely to play in the future development of tourism.

Literature Review

A government's role in tourism is the outcome of its tourism policy formulation and implementation (C.M. Hall, 1994). Therefore, an analysis of tourism policy is the most straight-

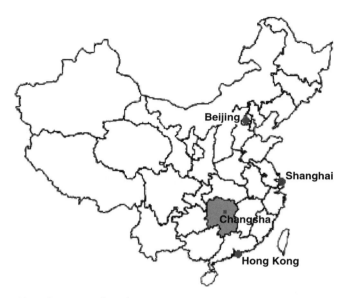

Fig. 12.1. The position of Hunan within China.

forward way to identify government roles in developing tourism. The role of policy, its interpretation and its implementation, has been described as a course of action calculated to achieve specific objectives (Brown and Essex, 1988: 533). It has also been viewed by other authors as either an agent of change or a response to change; and hence is dynamic. It can also be a product of other policies (Hill, 1993), being required to respond to issues and problems created by previous policies. In this way all policy is 'policy change' (Hogwood and Guy, 1983).

Richter (1983, 1989) earlier examined the evolution of Chinese tourism policy, its political environment and its organization, and found that it is critical to understand not only the design of the policy but the nature of the resources of the society and its administrative milieu. Edgell (1990) offered a model for the tourism policy-development decision-making process which integrates what Poon (1994: 35) refers to as 'frame conditions' (others may refer to the 'external environment'), goal objectives and resources (Burns and Holden, 1995: 186). According to C.M. Hall (1994), the tourism policy-making process includes:

- *policy demands*, formed both inside and outside the political system;

- *policy decisions*, taken by the political authority, which are authoritative rather than routine;
- *policy outputs*: what the system does; thus, for example, while goods and services (such as tourism) are the most tangible outputs, the concept is not restricted to this; and
- *policy outcomes (or impacts)*: the intended or unintended results from political action or inaction.

Mill and Morrison (1992) also identified five main areas of public sector involvement in tourism: coordination, planning, legislation and regulation, as entrepreneur, and stimulation. C.M. Hall (1994) added two other related functions: a social tourism role, and a broader role of interest protection. Yet although these authors have discussed the various roles that can be played by government, they do not address the nature of the interrelationship between these roles. A description of the roles only may provide a false impression that the various roles are independent from each other. This chapter attempts to address this shortcoming. Furthermore, most policy analytical frameworks take a Western perspective. Exceptionally, Hall (1991, 2001) suggested an evolutionary framework for state socialist countries, mainly referring to Eastern Europe and the former Soviet Union. However, most of these territories have since undergone dramatic political change. Whether his framework can also be used in China is still to be tested. Thus far there is no analytical framework available which is from a Chinese point of view, nor arguably, even from an Asian perspective (although see also Hall, 1990; Sofield and Li, 1998; Lew, 2001).

Research Methodology

Through broad exploratory study, a literature search, talking with experts and conducting a focus group interview, the author determined that C.M. Hall's (1994) tourism policy-making process model provided a useful loose framework for research. However, this chapter does not attempt to test empirically Hall's model nor examine the policy-making process in China. Rather, it relies on the use of a component of the policy-making model to guide an assessment and analysis of China's policies.

In order to provide a systematic analysis of Hunan tourism policies, the author conducted semi-structured interviews not only to reveal and understand the 'what' and the 'how' but also to place more emphasis on exploring the 'why' (Saunders et al., 2003: 245). The interviewees were:

- the provincial tourism officer;
- general managers of large Hunan hotels: FuRong – the first hotel in Hunan accommodating foreigners – and Huatain, the first 4-star and subsequently first 5-star hotel in Hunan;
- general managers of travel agencies, including Hunan International travel agency and smaller agencies;
- general managers at HuangHua Airport and Hunan rail bureau; and
- professors at Hunan Normal University, Hunan College of Finance and Economics and Hunan University.

The above interviewees had all experienced the three developmental periods discussed below and some were leaders in their field. The interviews were undertaken in the interviewees' native language (Chinese).

Documentary information such as maps, relevant policy documents and appropriate statistical data were collected from the National Tourism Office and statistical office, Hunan provincial tourism bureau and statistical bureau. The quality of tourism statistics in China, as elsewhere, is often criticized. Analysis may be limited particularly in regard to seasonal information. There are often bureaucratic delays and errors in compilation so that use and interpretation may be hazardous. System errors and statistical errors are quite common. However, in spite of many problems official figures can be helpful. Sometimes cross-referencing is possible, for example the Hunan provincial government tourism bureau reports tourists' spending in Hunan and these figures can be compared with those produced by other provincial government and China tourism departments. Provided that strict definitions are used, and data limitations are carefully and exactly reported, the regular series of international and national travel statistics can be put to practical marketing use.

Applying Hall's model to analyse collected data, the author sought to establish an holistic perspective of the government's roles during the three tourism development periods within Hunan and for the country as a whole. Based on these experiences, future government roles are also forecast.

The Provincial Government's Roles in Hunan Tourism Development

First period: 1978–1985

After the Cultural Revolution (1966–1976), the policy adopted in 1978 by the central government of 'opening doors to the outside world and enlivening the domestic economy' paved the way for rapid development in economy, culture, science and technology. Tourism development increased rapidly. At the same time the nature of tourism was changed from a mere political instrument to a legitimate commercial enterprise (Uysal et al., 1986: 113). At this time Deng Xiao-ping and Chen Yun became the supreme leaders of the Communist Party of China (CPC). Any policy change involving tourism matters, which traditionally served as a

political instrument, had to be approved by
Deng and Chen. Their attitude towards
tourism greatly influenced many policy
changes. Following the Cultural Revolution,
China faced backward economic conditions
and a serious shortage of capital. It was in
deep need of foreign exchange to finance further economic development activities, which
were being promulgated in an era of economic
reform. The importance of tourism as a means
of accumulating foreign exchange was recognized and emphasized by both Deng and
Chen, who delivered important speeches
about the economic benefits of tourism
(Zhang, 1995). These leaders both argued that
to develop tourism actively in China there was
a need to:

- enhance the friendship and unity between
  the people of China and the rest of the
  world;
- accumulate funds for modernization programmes; and
- satisfy the growing demand for economic
  security so as to achieve a satisfactory standard of living (Uysal et al., 1986: 113).

Deng emphasized that developing tourism
should first encourage those businesses which
could earn more money. Chen commented
that tourism was just like the export of scenic
spots, earning foreign exchange more rapidly
than the export of goods. They agreed that
greater effort was needed to develop tourism
(He, 1992: 57; Han, 1994a). Both Deng's and
Chen's attitudes toward tourism development
led to positive changes in tourism policy and
encouraged the shift away from tourism as a
political instrument to one of an economic
tool. In this period, the Chinese leaders
regarded the nature of tourism as both political
and economic, to achieve a 'double harvest'
(Han, 1994a). However, the condition of the
administrative structure and tourism infrastructure and facilities at this time could not cope,
mainly because demand outstripped the supply of facilities and services.

Demand for tourism policies

Increased tourism demand thus highlighted a
number of shortcomings in tourism management.

- *Ineffective tourism administration.* Since
  1978, Hunan tourism administration had
  seen the Provincial Bureau of Travel and
  Tourism (PBTT) amalgamated with provincial travel services. With a combining of
  PBTT and the Provincial Foreign Affairs
  office in 1985, PBTT became directly
  attached to Hunan provincial government.
  In this area of tourism administration, a policy was introduced which called for the
  combination of both government and busi-
  ness/enterprise functions, but the organization was unable to pursue the two functions
  effectively.
- *Ineffective management.* As in every industry in Hunan at the time, the manager's
  authority in tourism enterprises was
  restricted. China had adopted a planned
  economy in 1949 and resource allocation
  was fully controlled by the central government with managers having little autonomy. Due to the traditional political
  command structure, branch secretaries of
  the Communist Party of China – rather
  than managers – were responsible for
  operations. These factors led to decision
  making not necessarily based on economic
  rationality.
- *Insufficient tourism infrastructure and facilities.* The Cultural Revolution (1966–1976)
  had rendered tourism development a non-
  issue (Uysal et al., 1986: 114). Upon
  opening China's doors to international
  tourists, the capacities of hotels, travel
  agencies and civil aviation could not meet
  the demands of a large influx of visitors.
  Visitor arrival numbers in Hunan in 1978,
  for example, were just 6400, yet these
  exceeded the total accumulated arrivals
  from 1949 to 1977. In the following year
  (1979) this figure increased to 8100 visitors. However, at that time Hunan had
  only limited bed and breakfast provision
  for international tourists, the number of
  establishments rising slowly to 50 by 1979.
  The annual hotel growth rate of 9.4% was
  totally inadequate in meeting the basic
  accommodation needs of this surge of
  tourist arrivals (HNTA, 1985–1998; Qu,
  1999).
- *Non-profit-oriented tourism pricing.* Before
  1978, as tourism was primarily promoted

for political purposes, prices were controlled by the central government and merely set to cover operating costs; provinces such as Hunan had no authority to set new prices (Qu, 1996). These prices were fixed and the government was ignorant of demand fluctuations occurring between the different tourist seasons.

- *Poor service quality.* As far as tourism industry employees were concerned, like every one else on the government payroll, they were entitled to the benefits of the 'iron rice bowl': their jobs were safe and evaluation was not based on performance. Thus there were no incentives for employees to pay attention to service quality or its improvement. For example, Qu (1996) found that most tour guides did not provide the necessary interpretation on visits to famous places or to historic sites, but rather they would take a nap on the bus or wait in the museum gift shop. Of course, this was far from acceptable international standards and left visitors dissatisfied.

Policy decisions and outputs

SEPARATING ENTERPRISE FUNCTIONS FROM GOVERNMENT FUNCTIONS. As mentioned above, before 1978 the PBTT was under the jurisdiction of the Department of Foreign Affairs rather than the provincial government. In 1978, the State Council upgraded the status of BTT to the State General Administration of Travel and Tourism (SGATT) which came directly under its jurisdiction and SGATT became the sole government body responsible for tourism administration. Meanwhile, Hunan Province established its own tourism bureau. Later, in 1982, the State Council separated China International Travel Services' (CITS) enterprise functions from SGATT (He, 1992), and tourism administration became a government function that would no longer be involved in enterprise activities. At the same time, SGATT was renamed the China National Tourism Administration (CNTA). Because central government departments changed their structures and policies, provincial governments followed and changed as well, although Hunan Province only completed the adjustment in 1985 since tourism was not a priority at that time.

FROM CENTRALIZATION TO DECENTRALIZATION. Prior to 1978, civil aviation, travel agencies and hotels were funded and operated by central government. In 1984, the State Council decided that central and local government, collectives and even individuals could invest in and operate tourism development projects (Han, 1994a). The Civil Aviation Administration of China (CAAC) also encouraged local governments and individual government departments to operate airlines. From 1978 to 1985 there was just one dual-function airport for both military and civilian traffic in Hunan, which was funded and operated by the central government. Changsha airport was used by eight intra-provincial airlines and handled more than 20 flights per week (Qu, 1996). From 1985, with encouragement from the state council, Hunan planned to build a further two airports which the provincial government would fund and operate (Qu, 1996).

Originally, only the offices of CITS and CTS (China Travel Services) had authority to contact foreign tour operators. This centralized policy restricted promotion of market growth until in 1980 decentralization of marketing promotion to local offices was introduced. It provided an opportunity for Hunan to establish an international reputation. However, lack of integrated control and coordination resulted in overbooking, unexpected delays and unsatisfactory itinerary changes. In response to these problems, the State Council further modified its decentralization policy in 1984 and allowed all regional branch offices to contact and sell to foreign tour operators, and to notify the visa-issuing authorities. However, their sales plans were still subject to the approval of CITS (Han, 1994a). Prior to 1984, CITS, CTS and CYTS (China Youth Travel Service, set up in 1980), monopolized travel service operations through their branches in different provinces. In 1984, the State Council decided that travel agencies could be owned privately as well as collectively, and CNTA classified all agencies into three categories. Category one travel agencies were allowed to negotiate directly with foreign tour operators while the other two were restricted to arranging tour-related activities for foreign tourists coordinated by category one travel agencies. The effects of decentralization and liberalization saw the number of travel agencies and tourists received by travel agencies increase (HNTA, 1985–1998).

The decentralization policy also stimulated domestically owned hotel development, and many central government departments, local governments, government enterprises and even individuals participated in hotel construction.

REFORM OF TOURISM PRICING. In this period, tourism pricing was directed at making a profit and tourism prices increased annually. The provincial government became aware of seasonal demand fluctuations, and in the 1985 peak, shoulder and slack tourist seasons were identified, where prices for the peak season would be 6% higher than those of the shoulder season (Hu, 1994).

ENTERPRISE REFORM. Adoption of corporate management was first initiated by Deng Xiao-ping. It meant:

- separation of tourism enterprises from administrative bodies where autonomy on personnel, finance and operational matters was granted to a certain degree from higher administration;
- 'manager command' rather than 'Party secretary command', where the manager replaced the Party secretary with the responsibility for overall management; and
- rewarding staff based on their performance (Han, 1994a).

All tourism corporations in Hunan province – that is state-owned enterprises, including travel agencies and hotels – were also permitted to establish a strict appraisal system for staff. Rewards were based on their performance and staff with poor performance could be disciplined and even dismissed. These corporate management measures were in line with Western hotel management concepts; they accompanied and were stimulated by foreign investment. For sectors such as travel agencies and airlines, reforms were less effective, as political command and 'iron rice bowl' privileges were still deeply rooted here.

TOURISM EDUCATION AND TRAINING. Due to problems of ineffective management and poor service quality, emphasis on tourism education and training was now stressed by central government. As tourism served both political and economic requirements, a 'red and professional' education and training policy was adopted which aimed at developing tourism employees with the ideology of communism and socialism ('red'), but complemented with knowledge about tourism management and operation ('professional'). Between 1978 and 1983, when insufficient funds were available to establish educational tourism institutes and there was an urgent need for manpower, the HNTA offered financial and human resources to run tourism programmes jointly with selected universities. CNTA had established its own educational institutes, such that by 1986, a tourism education system was in place which comprised 189 vocational schools, four secondary professional schools and ten colleges and universities offering tourism courses (Zhang, 1987). As a result, many Hunan managers were sent to other provinces, and selectively sent overseas, for tourism management training.

Impact of tourism policies

The goal to earn more foreign exchange was achieved, with tourism receipts in Hunan increasing from US$0.82 million in 1978 to US$5.12 million by 1985: a more than sixfold increase. The share of tourism receipts in total foreign exchange earnings increased from 2.7% in 1978 to 4.6% in 1985, and the earnings growth rate from tourism increased faster than other 'exports' (HNTA, 1985–1998).

The unintended consequences that resulted from the Hunan provincial government's overemphasis on the quantity of tourism rather than its quality was that the policy allowed the introduction of foreign investment and decentralization too rapidly. Without proper planning, the consequent lack of coordination and central control resulted in what the Chinese called 'disorder tourism'.

From 1985 to 1988 the growth rate of hotels, hotel rooms and hotel beds all exceeded the main market demand for hotels. As a result, hotel supply moved ahead of the availability of transportation, especially civil aviation (Tisdell and Wen, 1991).

Although travel agencies in China had been classified into three categories, their delineation was not clear and was based solely on whether agencies were allowed to make foreign contacts. Violation of the regulations occurred and service quality remained poor; tourist guides remained frequent targets of complaints for their unacceptable level of service and questionable ethical practices.

During this period the economic contribution of tourism was recognized and led to the opening up of Hunan province to foreign tourists. Understandably, the initial influx of 6400 visitors in 1978 created carrying capacity problems, as government-owned or -operated facilities and services were hardly able to cope with the expectations and demands of the international traveller. A lack of infrastructure, facilities and poor service quality, coupled to ineffective administration and management, led the government to introduce policies designed to separate the enterprise/operational functions of government from its administration functions. To overcome and resolve the problems, the government introduced tourism pricing and enterprise reforms, and established education and training programmes. In this way, the government played the roles of implementer, investment stimulator, regulator and educator, respectively. This period was also characterized by a lack of coordination and a mismatch between luxury-class hotel provision and visitor needs. The provincial government clearly recognized the need to increase the supply of facilities and services and to upgrade the quality of service, and it acted accordingly. However, a lack of coordination and the unintended consequences of the hotel mismatch were acknowledged and ushered the next period, from 1986 to 1991, when a national tourism plan was adopted to guide the further development and prosperity of tourism.

Second period: 1986–1991

In December 1985, the government incorporated tourism into the Seventh Five-year National Plan as a key component for economic and social development. Tourism was declared to be a comprehensive economic activity with the direct purpose of earning foreign exchange for China's modernization (Han, 1994a; Zhang, 1995). This event was a significant benchmark for central government attitudes towards the role of tourism and saw a change in emphasis from a balance of the political and economic, to the economic having dominance over the political.

Demand for tourism policies

In developing tourism as an economic priority and resolving the problems of 'disorder' experienced during the first period, the China National Tourism Administration (CNTA) prepared a national tourism plan for 1986–2002. The government now paid full attention to the development of aviation, visitor attractions, tourism education, improvement of service quality, and the operation, promotion and coordination of tourism development. The plan focused on 21 top tourist cities, including Beijing, Shanghai and Hunan's capital, Changsha (Han, 1994a).

Policy decisions and outputs

COORDINATION AT GOVERNMENT LEVEL. Minimal coordination had led to disorder in the tourism industry, and the National Tourism Commission (NTC) was set up by the State Council in 1988 to assume responsibility for coordinating activities. Hunan Provincial Tourism Bureau took the decision to respond to central government policy. The LinLing area set up a tourism cooperation team to improve the area's tourism management; the NanYue area was concerned with attraction management, and different departments cooperated to improve visitor management at attractions (HNTA, 1985–1998). Both areas are rich in tourism resources and well known throughout China.

MASS RESTORATION AND DEVELOPMENT OF TOURIST ATTRACTIONS. During this period, the main emphasis of the Hunan government was on the provision of tourism infrastructure and facilities. At the National Tourism Conference in 1984, awareness was raised of the equal importance of developing attractions and increasing the carrying capacities of tourist facilities and services. As a consequence, Hunan tourism invested 10.75 million RMB (US$1.3 million), focusing on 12 major projects (HNTA, 1985–1998).

AVIATION REFORM AND AIRPORT DEVELOPMENT. Structural reform of CAAC was undertaken in 1987 with the aim of transforming airlines and airports into independent corporations. Airlines which now provided both international and domestic services were Air China, China Eastern Airlines, China Southern Airlines, China South-west Airlines, China North-west Airlines and China Northern Airlines. These would be responsible for their own profits and losses. Airport operation and management was transferred to local and regional governments (Han, 1994b). Hunan reconstructed Doumu Lake Airport in Changde (an important economic centre), and established a second-level airport in Zhangjiajie. At the same time, services between Changsha and Hong Kong, the major source of tourists, increased in number.

INTENSIFYING TOURISM EDUCATION AND TRAINING. During this period, Hunan provincial tourism bureau emphasized education and training for tourism. Until 1991, when the Taohuajiang women's tourism school was opened, there were just six tourism schools in Hunan (CNTA, 1985–1997). At the same time, different tourism departments educated persons from hotels, transport and travel agents, and helped to identify the required service quality standard. Forty-nine persons gained qualifications for the national tour guide examination in Hunan in March 1991.

REGULATING THE TOURISM INDUSTRY. With complaints of poor service quality and the deteriorating ethical conduct of some tour guides, Hunan Provincial Tourism Bureau followed CNTA in 1987 and 1988 with directives aimed at regulating all tourist guides over matters pertaining to their qualifications, duties and responsibilities. Tourist guides were required to be licensed and had to pass an examination before being granted such a licence. Service quality improved, but only to a limited extent, since the 'iron rice bowl' mentality was still deeply rooted in many tourism employees.

All hotels had to obtain a licence from CNTA as a means of reducing the mismatch between service quality and the grading of facilities, and to enable China to publicize its first group of star-ranked hotels, with 5-stars representing the highest standard (Yu, 1995).

Hunan achieved three 3-star hotel rankings, seven 2-star and nine 1-star rankings. Since then, hotel operations in Hunan have gradually improved to meet international standards, and 'disorder' problems within the industry have been reduced.

PROMOTING INTERNATIONAL TOURISM. With the increasing available capacity of tourism facilities and services, the Hunan provincial government began serious market promotion, and Hunan Provincial Tourism Bureau took advantage of the hosting of major events to attract international tourists. In 1991 for example, there were three international events: the Miluojiang dragon boat festival, the Hunan firework festival and the Zhangjiajie international forest conservation festival. The HNTA used the different category travel agencies in China and also international exhibitions to introduce Hunan to the world, and invited overseas media to the province.

IMPACT OF TOURISM POLICIES. These policies did not achieve their objectives, however, largely as a result of the negative publicity and imagery conveyed by the world's media following student deaths in front of PLA tanks in Beijing's Tiananmen Square on 4 June 1989. Neither tourism arrivals nor tourism receipts achieved their targets in Hunan. Moreover, poor service quality was still a persistent problem, although it had improved in foreign-owned and -managed hotels. According to a survey conducted by CNTA (Sun, 1989), one quarter of all Japanese tourists were dissatisfied with the overall quality of service in China.

In summary, this second period saw the introduction of policies designed to overcome the problems of 'disorder'. The government assumed the new roles of coordinator and planner through the establishment of a provincial tourism commission, and the adoption of a tourism plan, respectively. In the meantime, it continued to intensify its activities as an operator, regulator and educator. In terms of the latter role, the government had established a comprehensive and extensive tourism education system to accommodate the manpower and training needs of the industry. This period also saw the need for, and recognition of the importance of, international marketing promotion. Interestingly, this was not initially regarded as a key aspect of government responsibility in tourism. However, with the slowdown of tourist arrivals in 1987 and 1988, followed by a dramatic drop in arrivals after the events of 4 June 1989, effort was intensified and the budget for marketing promotion was increased. By 1991 tourism had developed into a significant industry and it was clearly recognized as an important economic activity in Hunan: there were now 38 hotels in the province which directly employed 50,563 staff (CNTA, 1985–1997).

Third period: 1992–present

In the spring of 1992, Deng Xiao-ping announced the speeding-up and intensification of economic reforms during his tour of Guangdong province. During the 14th Communist Party Congress in October 1992, a milestone resolution was adopted to pursue the policy of establishing a 'market economy under socialism' which would allow the market to determine resource allocation within the guidelines of socialism (Liu, 1993). This signified that China would open its door wider than before, and its economy would be further geared to the market system.

Demands for tourism policies

As an important economic activity, tourism would undoubtedly be closely tied to this general market-oriented policy, and be further opened to foreign investors. This also gradually geared Hunan to receive international tourists and to promote the overseas travel of Chinese nationals.

During the first 10 years of China's economic reforms, foreign revenue was badly needed to support the economy, and the Chinese government took measures to attract international tourists, while paying little attention to the development of domestic tourism. Despite these circumstances, domestic tourism was increasing because of the fast-growing economy and marked improvement in living standards. In addition, the Chinese nation has an age-old tradition of tourism, which emphasizes expanding one's knowledge, raising one's understanding of the world and society, enhancing self-cultivation, making friends and conducting cultural exchanges. All these have contributed to the rapid growth of domestic tourism (Zhang, 1995). As a government official recognized: 'domestic tourism should now be the foundation of China tourism' (Liu, 1995: 14).

Policy decisions and outputs

EXPANDING THE AREAS OF FOREIGN INVESTMENT. Prior to 1992, foreign investment was only allowed in the hotel sector, but with a more open economic reform policy, travel agency and aviation operators were no longer restricted. For example, of the US$3 billion investment required for the development of Yueyang holiday centre, US$2.7 billion came from foreign sources. Another example is the Hong Kong Anda international company's investments of US$10 million in the Tianzishan Chain Bridge in Zhangjiajie National Forest Park. The road between the Yellow Lion Village chain bridge and Suoxiyu was also built with foreign investment. The natural attractions to which these roads provide access are situated in the western part of Hunan in Zhangjiajie National Forest Park, which was the first national forest park in China, and declared a World Heritage Site by UNESCO in 1992 (UNESCO, 2001).

PLANNING OF RESORT CONSTRUCTION. Prior to 1992, vacation travel in China traditionally comprised sightseeing at natural and man-made scenic spots. In making tourist attractions more adaptable to the needs of the international market, the State Council decided in 1992 to build resorts to combine simple sightseeing with holidaymaking. Resort construction would employ both state and foreign investment. Following this policy, under PTB's plan, a number of holiday centres were developed in Changsha, in Yueyang and elsewhere.

The provincial government also offered attractive incentives for foreign investment in the state resorts, including a 24% tax allowance. For tourism corporations which had operated for more than 10 years, the Government would not impose any tax in the first and second years, and would impose half of the normal tax from the third to the fifth year (Qu, 1999).

DECENTRALIZATION OF TOURISM PRICING. In developing tourism based on market mechanisms, decisions on tourism pricing were devolved to individual tourism corporations in 1992. By 1994, all tourism enterprises were operating in a market economy environment, responding to international tourism market demand (Hu, 1994; Liu, 1995).

REGULATING THE TOURISM INDUSTRY. Regulations for tourism management in Hunan were issued in 1997, and stimulated the PTB to pursue a number of methods to improve management. These included conducting the first training classes for tourism regulation enforcement and supervisory staff (CNTA, 1985–1997).

By the end of 1997 the province had 167 travel agencies, of which 25 were international and 142 domestic. Focusing on the widespread problem of overcharging tourists for services, the PTB both emphasized ethical education and training, and strengthened enforcement and supervision, with strong punishment for illegal behaviour (HNTA, 1985–1998).

Impact of tourism policies

Tourism during the third period fully recovered from the dramatic drop in visitor arrivals experienced in 1989, and then developed rapidly. In 1998 there were 348,600 international tourist arrivals in Hunan and receipts from them reached US$15.6 million (HNTA, 1985–1998). It is generally considered that these achievements occurred largely as a result of mass promotional campaigns. Also important was the fact that Hunan's tourism industry had gradually geared itself to the free market economy environment, and had further expanded to 186 hotels and three airports (HNTA, 1985–1998).

In this third and current period, the government's role as investment stimulator, promoter and regulator has intensified. The regulation of tourism pricing has been relaxed so that tourism corporations can determine it themselves. The role of planners has been extended from the provincial tourism plan to involve-

ment in the establishment of state resorts as specific tourism products. A clear shift in policy has occurred in this period whereby a socialist market economic approach with Chinese characteristics has been adopted.

The demands of tourism policy making in Hunan derived more from within the political system in the first and second periods, but in the third period exogenous forces have become more important. During the analysis, for convenience, the author has combined the policy decision and policy outputs components which are separated in Hall's model. Hunan tourism policy in all three periods has been authoritative, and all the provincial government's objectives have been achieved. Unplanned impacts appeared in all periods, but these were largely rectified or eliminated as the result of subsequent policy decisions.

The provincial government's roles in the future

According to the World Tourism Organization's (WTO) 2020 vision report, China will be the number one tourism destination and the fourth ranked tourist-generating country in the world by 2020, with 137 million international visitor arrivals and 100 million outbound tourists (WTO, 1997). Building on its resources and tourism development experience, by the beginning of the 21st century, Hunan's tourism development rate was above the national average and had become one of China's ten leading provinces in this respect (Qu, 1999).

For the 21st century, Wei (1999) has identified that China faces eight main challenges in its tourism development. The present author forecasts that the following will be required in order to face these challenges in the tourism development of Hunan.

Improved capitalization of China's strength

The first challenge is improved capitalization on Hunan's strength in expanding tourism. As the promoter, that is possible only if HNTA has a clear idea of what this actually is: its long history, diversified culture and exquisite natural beauty. For most foreign visitors, it is the ancient history and diversified culture, as represented by the Xihan female corpse in Mawangdui, the dragon festival and the ethnic diversity of Xiangxi, which attracts them to Hunan. For most overseas Chinese, it is particularly the unequalled natural beauty found in the Zhangjiajie National Forest Park and the Heng Mountains.

Market research and product development

The second challenge for China in the new millennium is to do a better job in research and development, which means market research and new product development. For Hunan this is a new challenge, partly because of limited financial resources, and partly because of limited capability (Qu, 1999). In the past 20 years, when rapid tourism growth was achieved in Hunan, no large-scale research project was undertaken or commissioned by HNTA or any travel business in Hunan on any aspect of Hunan's major overseas markets. That may give the false impression that market research is not essential to the growth of tourism in Hunan, a mistaken impression that should be dispelled. The HNTA should also initiate proper research into the domestic tourist market.

In terms of new tourism product development, it is essential for Hunan not just to be a follower – doing things that others are doing without knowing why. In tourism we have seen in recent years the popularity of man-made attractions, holiday resorts, hotel construction and festivals (Wei, 1999). Therefore, the provincial government should act as the researcher not only into markets but also for new and improved tourism product development.

Keeping pace with technological advances

A third challenge facing Chinese tourism in the new millennium is to keep pace with technological advances. As the operators, it is essential for HNTA to be fully aware of the impact that technology will have on hotel and tourism products. As Cook *et al.* (1998: 387) have argued: 'Technology will become more important to service providers as rising wages force cuts in staff size, creating the need for increased productivity'.

Consolidating tourism businesses

The fourth challenge for China's tourism industry is to consolidate its operational structure – notably hotels and travel agencies – through mergers and acquisitions, or through forming closely knit, operational business groups.

It is universally acknowledged that the hotel industry is a fragmented one, but the situation in Hunan has been far worse than many people in the industry care to admit (Qu, 1999). As for Chinese travel agencies, they are small, weak and fragmented, as top CNTA officials admit (He, 1998). The world trend, however, is toward consolidation and acquisition. With China joining the World Trade Organization, foreign service providers will be able to establish themselves in the country and compete with Chinese companies. It is unlikely that a highly fragmented industry could compete against international chains armed with a digital 'nervous system' and abundant marketing expertise.

Therefore the HNTA should learn from the education department, which stimulates the universities to consolidate horizontally and vertically to improve the province's competitive power in the country and globally. It is also necessary to use the administrative power of the government intelligently in terms of rationalizing the organization and employing procedures recognized by the rest of the world.

Strategic cooperation with other Asian destinations

The fifth challenge that China faces is to enter into competitive or strategic cooperation with Hong Kong, Singapore, Thailand, Malaysia and other Asian countries. Hunan's main visitor markets are Hong Kong, Taiwan, Macao and other South-east Asian countries. However, Hunan is geographically in the middle of China and is a distant destination for its main markets compared to China's coastal provinces. Europeans and North Americans are not frequent visitors unless they are business travellers. The HNTA can respond to these issues by broadening its coordination and cooperation with other central Chinese provinces. This does not mean that individual destinations should give up their own distinctive features. The very opposite is true: the HNTA should coordinate with other provinces to develop complementary distinctive products for domestic tourist markets.

Intensified and continuous tourism education and training

The sixth challenge for China's tourism development is to improve considerably its tourism training and education. For Hunan, this is very important. There are now 18 tourism schools and colleges in the province, but the overall level is still low because 11 of the 18 are vocational schools. As an educator, the HNTA should emphasize the tourism teacher's education first. As far as school education is concerned, teachers of hotel and tourism management should be encouraged to get involved in the operation and management of hotels, travel agencies and other tourism businesses on the one hand, and follow world development closely on the other. The fact that teachers cannot read any foreign literature on travel and tourism because of their low foreign language competence must be changed as soon as possible. As an educator, the HNTA also should place adult and/or continuing education as a top priority for all those already involved in tourism, especially those holding leading positions in tourism administration and managerial positions in travel businesses. They should be kept up to date on changes and trends, new technologies and opportunities in the industry. HNTA should find an educational partner in tourism either in China or in another country so that a large group of tourism professionals can rapidly come to the fore. Also, academic exchanges between Hunan colleges of tourism and those of other provinces or foreign countries should be vigorously promoted.

Enhancing service quality

The seventh challenge China faces is to enhance service quality. In Hunan, although a large number of rules have been issued to regulate the industry and improve service quality, some have not been effective as enforcement continues not to be efficient. This equally applies to other roles. For example, most receptionists in Hunan hotels are not good at foreign languages and may not understand foreign languages as a result of their tourism teachers' low foreign-language competence.

In the face of this challenge, the HNTA should emphasize the training of enforcement personnel and should better coordinate their roles.

Privatizing part of the travel businesses

A large number of joint-venture hotels have been playing a significant role in Hunan's tourism expansion, although the majority of travel businesses in the country are publicly owned and operated. The private travel sector is very small. As Premier Zhu Rongji stated in his 5 March 1999 government report:

> Active steps should be taken to encourage, support and guide individual economy and private economy and other types of non-public-ownership economy to grow healthily so that their role can be brought into full play in meeting the diversified needs of the people, providing jobs and promoting the development of the national economy.
>
> (Wei, 1999: 28)

Therefore, the HNTA will acquire a new role to encourage, support and guide the private sector economy. Partly this is because privatization can alleviate some of the chronic problems Hunan faces in its tourism development. It is true that privatization will not eliminate all wrong decision making, but it is likely that fewer poor major investment decisions will be repeated, and fewer resources will be wasted.

The above are suggestions for government roles in Hunan tourism development based on Wei's (1999) critique of China as a whole. As Coccossis (1996) has argued, an important characteristic of the interaction between tourism and the environment is the existence of strong feedback mechanisms: while tourism often has adverse effects on the quantity – and quality – of natural and cultural resources, it is also affected by the decline in quality and quantity of such resources. The provincial government must therefore pay attention to the natural and cultural base upon which tourism depends. Globally, 'sustainability' has only been recognized explicitly in the last 20 or 30 years (Swarbrooke, 1999): central government in China incorporated this concept into the national development plan and has emphasized it since 1992. The provincial government, as a regulator, should also regulate the industry and influence tourists to foster 'sustainable tourism', rather than waiting for serious negative impacts to appear and then regret their inaction.

Conclusions and Recommendations

Hunan is rich in natural and cultural tourism resources. During the past two decades, when China opened its doors to international tourists, tourism in Hunan developed rapidly and became a significant economic sector. The role of tourism has been transformed from being initially a political tool, centrally controlled, to an economic one, now driven by market forces.

Given the nature of China's economic development under communist rule with strong central government control, it is not surprising to find that Hunan provincial government has played a key and decisive role in shaping the development of tourism through the adoption of a specific series of policies. For each of the periods identified, the roles played by the government and the accompanying policies were quite distinct and pivotal in shaping the nature and direction of tourism. The various roles played by the provincial government are interrelated, and, if effective, should bring about a 'synergistic effect' to achieve the intended goals for tourism.

Hunan's experience in the development of international inbound tourism has demonstrated that it has been possible to develop tourism actively with limited private sector involvement. The provincial government, through its various roles, has been a major player contributing to the current success of the industry. Despite some unintended impacts, it has been able to overcome many of the problems resulting from rapid tourism development. Adaptability has been a key aspect of Hunan's tourism policy development and this needs to be sustained as governmental roles change and become more complex over time.

In conclusion, the following roles are identified for provincial government in order that they be able to respond to the challenges of further developing Hunan tourism:

- as a marshal, carrying out the national tourism office decisions consistently and realistically;
- as a regulator, paying attention to environmental protection and taxation and fostering the growth of Hunan's tourism in a sustainable way;
- as a coordinator, not only coordinating activities of different government departments with respect to tourism, but also cooperating with other central provinces to promote tourism in Hunan and the wider central Chinese region;
- as a promoter, better capitalizing on its strength in expanding tourism;
- as a researcher, not only undertaking sound market research for its markets but also for its new products;
- as a stimulator, not only encouraging foreign investment or social funds, but also stimulating, supporting and guiding tourism in the private sector; and
- as an educator, not only providing training courses but also finding education partners from other provinces or foreign countries so that a large group of tourism professionals can quickly be brought to the fore. Education and training for tourism teachers should also be a top priority.

Tourism in Hunan has certainly come a long way from when it was based on a centrally planned economy to a position where it is able to respond to market needs. Time and continued experience with the market should enable Hunan tourism to flourish in the future.

References

Brown, G.P. and Essex, S.J. (1988) Tourism policies in the public sector. *Journal of Travel Research* 26, 533–539.

Burns, P.M. and Holden, A. (1995) *Tourism: a New Perspective*. Prentice-Hall, London.

CNTA (China National Tourism Administration) (1985–2001) *The Yearbook of China Tourism Statistics*. China Tourism Publishing House, Beijing.

CNTA (China National Tourism Administration) (2003) *Welcome to China: China Tourism: Hunan Summary*. CNTA, Beijing <http://www.cnta.gov.cn>.

Coccossis, H. (1996) Tourism and sustainability: perspectives and implications. In: Priestley, G.H., Edwards, J. and Coccossis, H. (eds) *Sustainable Tourism? European Experiences*. CAB International, Wallingford, UK, pp. 1–21.

Cook, R.A., Yale, I.J. and Marqua, J.J. (1998) *Tourism – the Business of Travel*. Prentice-Hall, London.

Edgell, D.L. (1990) *International Tourism Policy*. Van Nostrand Reinhold, New York.

Hall, C.M. (1994) *Tourism and Politics: Policy, Power and Place*. John Wiley & Sons, Chichester, UK.

Hall, D. (1990) Stalinism and tourism: a study of Albania and North Korea. *Annals of Tourism Research* 17(1), 36–54.

Hall, D. (ed.) (1991) *Tourism and Economic Development in Eastern Europe and the Soviet Union*. Belhaven, London.

Hall, D. (2001) Tourism in communist and post-communist societies. In: Harrison, D. (ed.) *Tourism and the Less Developed World: Issues and Case Studies*. CAB International, Wallingford, UK, pp. 91–107.

Han, K.H. (1994a) *China: Tourism Industry*. Modern China Press, Beijing.

Han, K.H. (1994b) *China: Civil Aviation Industry*. Modern China Press, Beijing.

He, G.W. (ed.) (1992) *China Tourism System Reform*. Dalian Press, Dalian.

He, G. (1998) Speech at the National Conference on Tourism. *China Tourism News* March, p. 12.

Hill, M. (1993) *Understanding Social Policy*. Blackwell, Oxford.

HNTA (Hunan Tourism Administration) (1985–1998) *Hunan Tourism Statistics*. Hunan Tourism Administration, Changsha.

Hogwood, B. and Guy, B.P. (1983) *Policy Dynamics*. Wheatsheaf Books, Brighton, UK.

Hu, D. (1994) China's quotation system to go international. *Travel China*, 25 September, p. 6.

Lew, A.A. (2001) Tourism development in China: the dilemma of bureaucratic decentralization and economic liberalization. In: Harrison, D. (ed.) *Tourism and the Less Developed World: Issues and Case Studies*. CAB International, Wallingford, UK, pp. 109–120.

Liu, C.X. (1993) Mixed diagnosis for mainland fever. *Free Business Review* 43–47.

Liu, C.X. (1995) Tourism administration goes macro. *Travel China*, 30 April, p.14.

Mill, R.C. and Morrison, A.M. (1992) *The Tourism System: An Introductory Text*, 2nd edn. Prentice-Hall International, Englewood Cliffs, New Jersey.

Poon, A. (1994) *Tourism, Technology and Competitive Strategies*. CAB International, Wallingford, UK.

Qu, F. (1996) *Hunan Tourism Market Study*. Hunan Geography Press, Changsha.

Qu, F. (1999) *Hunan Tourism Industry*. Hunan Geography Press, Changsha.

Richter, L.K. (1983) Political implications of Chinese tourism policy. *Annals of Tourism Research* 10(3), 395–413.

Richter, L.K. (1989) *The Politics of Tourism in Asia*. University of Hawaii Press, Honolulu, Hawaii.

Richter, L.K. and Richter, W. (1985) Policy choices in South Asian tourism development. *Annals of Tourism Research* 12, 201–217.

Saunders, M., Lewis, P. and Thornhill, A. (2003) *Research Methods for Business Students*, 3rd edn. Pearson, Harlow, UK.

Sofield, T.H.B. and Li, F.M.S. (1998) Tourism development and cultural policies in China. *Annals of Tourism Research* 25(2), 362–392.

Sun, S.C. (1989) *Tourism Economy Research of China*. People's Press, Beijing.

Swarbrooke, J. (1999) *Sustainable Tourism Management*. CAB International, Wallingford, UK.

Tisdell, C. and Wen, J. (1991) Foreign tourism as an element in PR China's economic development strategy. *Tourism Management* 12(1), 55–67.

UNESCO (2001) *Wulingyuan Scenic and Historic Interest Area*. UNESCO, Paris <http://whc.unesco.org/sites/640.htm>.

Uysal, M., Wei, L. and Reid, L.M. (1986) Development of international tourism in PR China. *Tourism Management* 7, 113–119.

Wei, Q. (1999) China tourism – new role for a new millennium. Paper presented at *Asia Pacific Tourism Association Fifth Annual Conference*, 23–25 August, Hong Kong.

WTO (World Tourism Organization) (1997) *The World's Most Important Destinations in 2020*. WTO, Madrid.

Yu, L. (1995) China's hotels landscape: a marker of economic development and social change. In: Lew, A. and Yu, L. (eds) *Tourism in China: Geographic, Political, and Economic Perspectives*. Westview Press, Boulder, Colorado, pp. 155–175.

Zhang, G.R. (1987) Tourism education in PR China. *Tourism Management* 8(3), 263–266.

Zhang, G.R. (1995) China's tourism since 1978: policies, experiences and lessons learned. In: Lew, A. and Yu, L. (eds) *Tourism in China: Geographic, Political, and Economic Perspectives*. Westview Press, Boulder, Colorado, pp. 3–17.

13 Authenticity, ethnicity and social transformation at World Heritage Sites: tourism, retailing and cultural change in Lijiang, China

Takayoshi Yamamura

Introduction

Many communities where World Heritage Sites (WHS) are located run the risk of being unable to cope with the social and cultural repercussions of the dramatic increase in tourists resulting from being listed as WHS (Martin, 2000). Importance has been attached to the commercial use of protected structures as a means of sustaining heritage (World Bank, 1998). Yet, it frequently happens that the inability of communities to take independent initiatives results in the injection of outside capital, which in turn drives the original local inhabitants out of historical areas (UNESCO Asia and Pacific Regional Bureau, 2000a). In response, UNESCO has pointed out the importance of the appropriate preservation of cultural heritages in their entirety, which entails not simply the physical preservation of sites, but also preservation of the local social and cultural structures and environments that constitute the settings of the sites in question. Hence, UNESCO is calling for independent initiatives on the part of local communities that will make such overall preservation possible (UNESCO Asia and Pacific Regional Bureau, 2000b).

Given such a background, this chapter addresses issues that the tourism industry faces by taking as its subject the World Heritage Site of the old town of Lijiang, Yunnan Province,

China, and by focusing on the tourist shops to be found there. For this purpose, ways were examined in which both the existing aboriginal society (a minority known as the *Naxi*), and the majority peoples coming in from outside, engage in commercial activities related to tourism. To this end, the study first focuses on business categories and operator attributes in order to clarify the actual conditions that have given rise to the increase in souvenir shops. In the light of this information, the study proceeds to investigate the structures whereby the native and incoming populations have been involved in the emerging prosperity of the tourist industry by examining the forms of shop occupancy and regional character of the items on sale. The information gained is then used to clarify issues faced by native communities as they engage in the development of tourism, in as much as it is these communities that should ideally be responsible for the management and perpetuation of cultural heritage.

Despite numerous discussions to date by UNESCO and other international bodies on the subject of relationships between the tourism industry and cities that are World Heritage Sites, there exist few cumulative data from specific studies based on on-site surveys. One of the most notable has been the study related to questions of heritage management and tourism, which was conducted by

UNESCO Asia and Pacific Regional Bureau (UNESCO Asia and Pacific Regional Bureau, 2000a,b). Although that report contained results related to issues pertaining to tourism in nine Asian World Heritage Sites, including Lijiang, no attempt was made to clarify the actual conditions prevailing in the regional tourist industry. The study itself went no further than a basic statistical survey and an outline of the issues involved.

In terms of architecture and urban planning, research has been initiated in recent years, notably by the International Council on Monuments and Sites (ICOMOS) and other organizations, into the repercussions of the creation of tourist sites as it affects World Heritage cities. Reports on such World Heritage cities as the old town of Lijiang in China and the historic centre of Telč in the Czech Republic (Drdácký, 2002) have emerged. Yet, although such research summarizes the issues arising from the repercussions of the tourist industry on both the buildings and spatial environment of the city/settlement, there has been no adequate survey or examination of either the structures involved in the increasing prosperity of regional communities and the tourist industry or of independent initiatives on the part of regional communities and native populations in managing their heritage.

While sharing an awareness of the issues entailed in such studies and research, this chapter takes an original approach in terms of its:

- specifically clarifying actual conditions of the tourism industry by using data from on-site surveys of the shops themselves as the basis of an examination of business categories and operator attributes; and
- focusing on the nature of the involvement of the various types of shop operators in the tourism industry in terms of regional characteristics, so as to examine problematic areas of the existing tourism industry in light of the interrelationship with the native community.

Additionally, it is felt that Lijiang is a suitable subject for a study involving the issues described above, as it is the historical home of the Naxi, who are heirs to a unique society and culture. Hence, the rise in the prosperity of the tourism industry is likely to affect the regional community and its culture to a marked extent.

Outline of Research Area and Methodology

Geographical outline of Lijiang

The Lijiang Naxi Autonomous County (hereinafter, Lijiang County) is located in the northwest corner of Yunnan province in south-west China (Fig. 13.1). Mountainous terrain covers 70%, and flat plains only 5%, of the land area. Lijiang County is about 600 km away from the provincial capital Kunming, and has derived comparatively little benefit from economic growth in coastal and central China.

The great majority of the population of Lijiang County are Naxi, an indigenous ethnic minority group, with the right of self-government (Fig. 13.2). The total population of the Naxi is about 278,000 across China as a whole and they retain their own language, pictographs and religion (Guo et al., 1999). There are approximately 198,000 Naxi living in Lijiang County, representing about 66.5% of the total Naxi population in China (Lijiang Prefecture Administration Office, 1997).

The Lijiang County seat consists of a new administrative section and the original town of Dayan. The old town of Dayan is known as 'the old town of Lijiang' or 'the ancient town of Lijiang' in which wooden buildings constructed in the Song (end of the 12th century to the early 13th century), the Yuan (early 13th to mid-14th century) and Ming dynasties (mid-14th to mid-17th century) are still standing (Fig. 13.3).

The old town of Lijiang was listed by UNESCO as a World Heritage Site in December 1997 in appreciation of the architecture of its dwellings and the historical townscape as a collective entity. The old town of Lijiang has a population of 14,477 housed in some 4156 wooden dwellings crowded into an area of 350.2 ha (figures provided by the Lijiang County authorities, March 2000). The whole area of the old town was designated as a conservation area by the government of Yunnan Province in 1997. The conservation designation plan places restrictions on the outward appearance and materials of buildings in this area. Extensions and remodelling of façades are strictly limited and new architectural designs are prohibited by guidelines stated in the plan (Government of Yunnan Province, 1997).

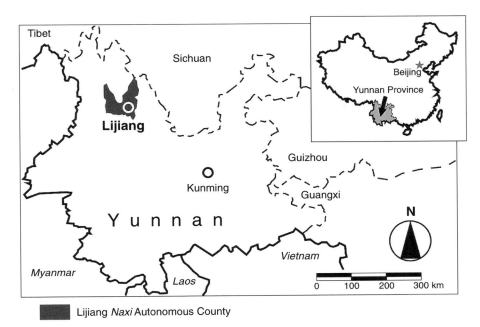

Lijiang *Naxi* Autonomous County

Fig. 13.1. Location of Lijiang (source: map courtesy of the author).

Fig. 13.2. The *Naxi*.

Methods

The survey sample consisted of all shops located in the central area of the old town (Fig. 13.4). This geographic area, approximately 230 m (north–south) by 200 m (east–west), is both the centre of the historic old town and an area of highest-priority conservation (56.775 ha), as designated in the conservation plan (Government of Yunnan Province, 1997).

This research area is also the central tourist focus, where tourists and shops are highly concentrated. Taking these geographical factors into consideration, it is believed that these characteristics support the suitability of the area for an on-site survey.

For purposes of this study, a single shop unit was defined as a section of a building in some manner demarcated from adjacent shops and being used by a single independent operator. This definition was adopted because premises were frequently encountered that were originally single units and were now partitioned into segments and conversely, that were originally multiple units but were now used as a continuum subsequent to the removal of dividing walls.

Fig. 13.3. Lijiang: view of the central area of the old town.

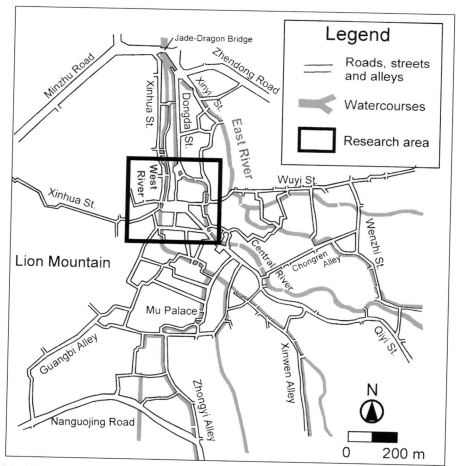

Fig. 13.4. Lijiang: research area (source: map courtesy of the author on the basis of the map by Xu, 2000).

The on-site survey was conducted between April and June 2000, personally by the author in accordance with the procedures described below.

- In China, detailed city maps are not available due to reasons of national defence. Therefore, with Lijiang County's permission, a map of the town was initially drawn up and prepared by the author. Some 286 shops were identified in the survey sector and constituted the subjects of the survey.
- The on-site survey was used to confirm details of the business categories and descriptions of the items on sale in the shops under survey.
- In light of such information, a person-to-person interview concerning attributes of the shop operators was conducted with operators of each of the 286 shops covered by the survey (operators themselves in self-run outlets; relevant information-desk staff in corporate outlets). During the interview, the principal information elicited on operator attributes included family register classification, race, place of birth, date of relocation and the date when the shop went into business. Questions relating to shop construction included ascertaining whether premises were owned or rented, and amount of rent paid. All interviews were conducted in Mandarin (standard Chinese) by the author. Each interview was based on a series of prepared questions.

With regard to such matters as social systems and fluctuations in the number of tourism-related industries, interviews with officials and residents were carried out to supplement locally available statistics and records, of which there is a severe lack.

Results

Classification of shops and their operators

Processes underlying increases in tourist shops

Table 13.1 shows the business categories of the shops and their respective numbers as determined by the on-site survey. Businesses were classified into two categories: 'tourist shops' and 'general shops'.

- Tourist shops: these targeted tourists as users of their services and purchasers of their goods. When multiple business activities are concurrently carried out, businesses catering for tourists as their main commercial activity are classified as tourist shops (Fig. 13.5).
- General shops: these targeted general residents as users of their services and purchasers of their goods. Alternatively, shops catering to an unspecified variety of customers, which could include tourists, such as outlets for foodstuff and general goods, hairdressers and other services, also belong in this category.

Further, according to the particulars of business content, businesses were sub-classified into categories of commodity sales/service industry, providers of food and drink, and establishments

Table 13.1. Lijiang shops' business categories.

Business categories	Number of shops	Percentage
Tourist shops	189	66.1
Commodity sales/service industry[a]	145	50.7
Providers of food and drink[b]	38	13.3
Accommodation[c]	6	2.1
General shops	97	33.9
Commodity sales/service industry[d]	88	30.8
Providers of food and drink[e]	9	3.1
Overall	286	100.0

[a] e.g. souvenir shop.
[b] e.g. restaurant, coffee shop, bar.
[c] e.g. guest-house, inn.
[d] e.g. general store, clothing store, barbershop.
[e] e.g. local eating house.

Fig. 13.5. Lijiang: an example of a tourist shop (a jade shop).

offering overnight accommodation. Figure 13.6 shows these results in map form.

Of all shops, 66.1% belonged to the tourism category. Although historical data show that the central quarter of the old town contained numerous shops in the past, most were apparently engaged in the sale of foodstuffs and daily provisions to local inhabitants (Goullart, 1955). The lack of surviving statistical records in the locality prevents our demonstrating from documentary evidence exactly when the business category of tourism began to increase in number. However, an examination of the dates when the shops went into business reveals that 176 (93.1%) of the 189 shops in the tourism category opened for business no earlier than 1995 (Fig. 13.7). Although the lack of records on the modes of use of the buildings prior to their being opened as shops leaves something to be desired in terms of infallibility of the evidence, it would seem reasonable to regard 1995 or thereabouts as the turning point, when the increase in the business category of tourism began for the following reasons:

- Lijiang airport came into operation in June 1995. Considerable improvements were made with regard to access to Kunming City, the capital of Yunnan Province. These developments were no doubt highly effective in attracting travellers, subsequently bringing about dramatic increases in the number of people journeying as far as Lijiang and in the income gained from tourism (Fig. 13.8); and
- the 'Conservation Detail Plan for Lijiang Old Town' formulated in 1997 not only provides regulations governing building façades, but also seeks to promote tourism by encouraging the modernization of building interiors and the commercial use thereof for purposes of tourism (Government of Yunnan Province, 1997).

Likewise, views corroborating this opinion emerged from interviews conducted by the author in June 2000 with ten local government officials in the county of Lijiang.

Moreover, the following two factors may account for the considerable number of shops opening between 1998 and 2000:

- Lijiang became more widely known as a result of having been listed as a World Heritage Site in December 1997; and
- as a result of the International Horticultural Exposition being staged in the city of Kunming in 1999, and Lijiang being designated as a subsidiary convention venue and included in the route of a tourist excursion from Kunming City, some visitors to this event also made their way to Lijiang, further boosting the number of travellers entering the town (Fig. 13.8).

These factors led to what might be justifiably described as a 'boom' in the opening of shops belonging to the 'tourist' category. This was confirmed by the information received from interviews with shop operators in June 2000, many replying that they viewed the staging of the horticultural convention as a business opportunity.

Shop operator attributes

Table 13.2 shows whether the shop operators surveyed are individual or corporate entities

Table 13.2. Lijiang: categories of shop operators.

Business categories	Categories of shop operators								
	Individual operators								
	Permanent inhabitants			Temporary residents			Corporate entities	Un-identified	Overall
	Naxi	Han	Others	Han	Naxi	Others			
Tourist shops									
Commodity sales/ service industry	48	6	1	70	1	15	4	0	145
Providers of food and drink	26	2	0	7	0	3	0	0	38
Accommodation	5	0	0	0	0	0	1	0	6
Sub-total	79	8	1	77	1	18	5	0	189
	41.8%	4.2%	0.5%	40.7%	0.5%	9.5%	2.6%	0.0%	100.0%
General shops									
Commodity sales/ service industry	42	5	2	17	0	4	15	3	88
Providers of food and drink	2	0	0	7	0	0	0	0	9
Sub-total	44	5	2	24	0	4	15	3	97
	45.4%	5.2%	2.1%	24.7%	0.0%	4.1%	15.5%	3.1%	100.0%
Overall	123	13	3	101	1	22	20	3	286
	43.0%	4.5%	1.0%	35.3%	0.3%	7.7%	7.0%	1.0%	100.0%

and classifies shops according to business category. In addition, the table gives the racial composition of the 263 individual operators, referring to their family register classification under Chinese law. Those with permanent residential qualification for Lijiang are classified as permanent inhabitants and those with temporary permits as temporary residents. Of the 286 shops surveyed, 263 (more than 90%) were run by individuals. Of these, 139 (48.6%) shops were operated by permanent inhabitants, mostly by Naxi, the aboriginal people of Lijiang, while 124 (43.4%) were run by temporary residents from outside the locality, principally Han, the major ethnic group of China. Turning to the tourism business category, it was found that, by race, it is the Naxi permanent residents that are the most numerous, at 41.8%. Yet the temporary residents are in the majority, totaling 50.8%. This and the fact that some 96 (77.4%) of the 124 operators with temporary-resident status are engaged in the tourism business category shows that the influx of population from outside the region is a factor that has been largely involved in the increase of tourism as a business in the locality.

Restrictions on population movement have been relaxed since the second half of the 1980s. Family registers are controlled by the Bureau of Public Security, and anyone wishing to reside in Lijiang must undergo screening and complete official documentation procedures at a local office controlled by the Bureau of Public Security before being issued a temporary permit, valid for a specified period of no longer than 1 year. Upon the expiration of this permit, an extension of up to 1 year is normally possible after a second official examination. The stringent legal restrictions on changing a registered permanent residential qualification make it extremely difficult for temporary residents to acquire permanent residential qualification in Lijiang. (Information gained from interviews conducted with Lijiang county Bureau of Public Security in June 2000 and Dayan Town Government in November 2000.)

Forms of shop ownership and the rented housing market

Shop ownership

Table 13.3 summarizes the conditions of building ownership for the 263 individually run shops, with figures being presented separately for permanent inhabitants and temporary

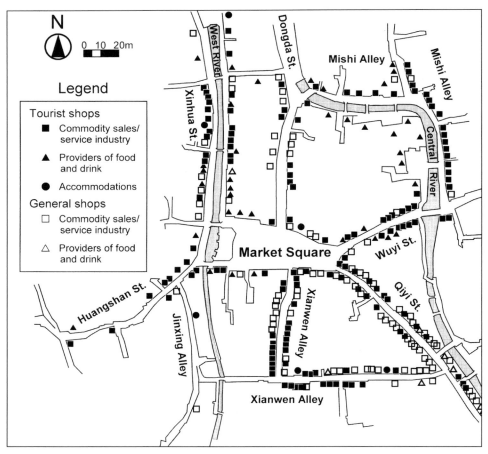

Fig. 13.6. Lijiang: current shop distribution.

residents. In the case of both permanent inhabitants and temporary residents, the majority of
operators used rented dwellings for their shop
premises. This was particularly so in the case of
temporary residents, with more than 60%
using premises rented from the private sector.
Such a market for rented housing has come
into existence since the 1980s as a result of the
Chinese government's efforts to promote provisions for a real-estate market as part of its
economic reforms. Such market developments
are currently evident throughout the nation.
Additionally, the interview survey revealed that
all of the owners of these private-sector rented
premises were *Naxi* with permanent-inhabitant
status, of whom 33.1% resided in the new
town, which lies adjacent to the old town
(Table 13.4). The new town has been under

construction since 1985 and is expected to
cover a projected area totalling approximately
15 sq km. By June 2000, approximately
20,000 people resided in the new town (statistical data provided by Government of Lijiang
County). According to a survey conducted by
the Dayan Town government, 5001 residents,
constituting 1527 households, had moved
from the old town to the new town between
1987 and 1999. This worked out at a relocation rate of approximately 32.7% of the old
town's population during that 13-year period
(figure calculated by Dayan Town government
in September 1999 from aggregated statistics
in possession of Residents' Committees in the
various districts under its control). Although,
strictly speaking, it would be necessary to ver-
ify this by conducting a survey of the actual

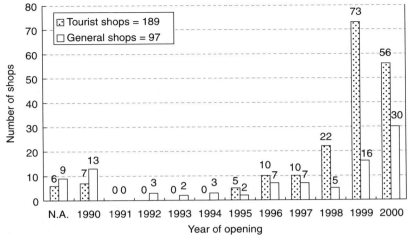

Fig. 13.7. Lijiang: opening-years of shops (source: compiled by author on the basis of information gained in interviews with shop operators).

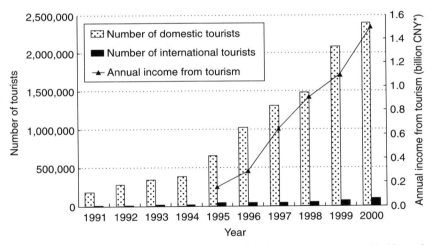

Fig. 13.8. Lijiang: trends in numbers of visitors and income from tourism (source: compiled by author on the basis of statistical data provided by the Bureau of Tourism, Government of Lijiang County).
Note 1: Figure for 2000 is the Bureau of Tourism's estimated value as of May 2000.
Note 2: Income from tourism refers to the total amount spent in the locality by tourists in the course of their journey.
Note 3: Statistical data indicating total income from tourism for 1994 and earlier have been omitted owing to lack of records in the locality.
*Note 4: RMB 1 (China Yuan Renminbi) = US$0.12.

owners, it can probably be inferred from the foregoing that many of the shops currently in use as tourism businesses were buildings rented subsequent to such relocation of the residents. This would suggest that it has been the availability of just such rental housing, primarily consisting of *Naxi*-owned housing offered for rent in the private sector, that has made it possible for temporary residents to make the kind of advances in the tourism business that were described in the foregoing section.

Table 13.3. Lijiang: ownership categories of 263 shops run by individuals.

Categories of individual shop-operators	Business categories	Number of shops	Ownership category				
				Rented houses			
			Own property	Private owner	Public owner	Company owner	Unidentified Others
Permanent inhabitants	Tourist shops	88 100.0%	29 33.0%	39 44.3%	17 19.3%	2 2.3%	1 1.1%
	General shops	51 100.0%	13 25.5%	22 43.1%	10 19.6%	6 11.8%	0 0.0%
	Sub-total	139 100.0%	42 30.2%	61 43.9%	27 19.4%	8 5.8%	1 0.7%
Temporary residents	Tourist shops	96 100.0%	2 2.1%	60 62.5%	23 24.0%	6 6.3%	5 5.2%
	General shops	28 100.0%	0 0.0%	18 64.3%	9 32.1%	0 0.0%	1 3.6%
	Sub-total	124 100.0%	2 1.6%	78 62.9%	32 25.8%	6 4.8%	6 4.8%
	Overall	263 100.0%	44 16.7%	139 52.9%	59 22.4%	14 5.3%	7 2.7%

Table 13.4. Lijiang: place of residence of private owners of rented premises.

Place of residence	Number of owners (= Number of shops)			
	Tourist shops	General shops	Total	Percentage
Lijiang County				
Old town	42	19	61	43.9
New town	31	15	46	33.1
Others	3	0	3	2.2
Sub-total	76	34	110	79.1
Other areas in Yunnan province	6	0	6	4.3
N.A.[a], others	17	6	23	16.5
Overall	99	40	139	100.0

Note: All owners were *Naxi* with the status of permanent inhabitant.
[a]Many tenants do not know where their owners live.

Rent charged for shops

Interviews were conducted to ascertain the monthly amount of rent paid on each of the 198 shops on premises rented in both the private and public sectors. Results revealed that the average rent for premises in the private sector and the public sector were, respectively, RMB1036 (US$125; median: RMB700) and RMB1766 (US$213; median: RMB1000). Valid responses in interviews were obtained from 127 (91%) of the 139 premises in the private sector and 50 (85%) of the 59 premises in the public sector.

Owing to the lack of locally available data concerning how such rental charges are related to factors such as floor-space, quality of living environment and location, the figures given above must be regarded as merely values for convenience of reference. None the less, these figures are based on the rental charges that applied whenever an individual rented a single shop building. Yet, the amounts appear quite high for a single individual to pay, especially when one considers that the GDP per person in Yunnan Province is RMB4355 (approx. US$527) (National Bureau of Statistics of China, 1999).

As noted above, guidelines are currently in force in the old town, governing the appearance resulting from reconstruction or repairs of existing buildings, although as yet there is no system of aid available to provide public funding for reconstruction and repairs. Hence, the owners of the buildings must incur all costs involved in such work (from interview survey conducted with old-town residents and Lijiang County government, May 2000). Further, although local real-estate-related data were not available, it is possible that land rents are rising steeply in accordance with market principles, given that the locality's listing as a World Heritage Site has increased the value of land as a tourist resource in the present-day context of an increasingly liberalized real-estate market. This factor may be related to the high rental charges of shop premises, although many aspects remain to be clarified. This issue deserves to be examined further.

Local characteristics and their proponents in shops

Types of souvenir shops in terms of local characteristics of products sold

Table 13.5 provides information on 145 shops selling tourist goods and services (hereinafter, tourist souvenir shops) that account for more than 70% of the business category of tourism in the locality surveyed, focusing on the presence or otherwise of local products in an attempt to classify them. Classifications included 'Lijiang specialities', i.e. goods, crafted items or other commodities produced or manufactured in Lijiang Prefecture and demonstrating in satisfactory measure the local characteristics of Lijiang. Other items in this classification include printed matter and the like generally available solely in Lijiang. Other categories include 'specialities from elsewhere', i.e. speciality products obviously produced in any locality other than Lijiang and being sold in Lijiang and 'products deficient in local characteristics', i.e. mass-produced goods also available elsewhere, possessing no specific regional characteristics. The table further provides a summary of operator attributes for each of these types of shop.

Little more than one-third of the shops sold Lijiang specialities, whereas specialities from elsewhere and products deficient in local characteristics were in greater evidence. In particular, the most numerous kind of shop was that which principally sold jade. There were 36 such shops, which represented 24.8% of tourist souvenir shops. Imported from Myanmar via the southern regions of Yunnan Province, these jade items have nothing whatsoever to do with the natural environment or culture of Lijiang. Further, a look at operator attributes reveals that just under 60% of all shops are run by temporary residents, and particularly in the case of specialities from elsewhere, it was found that 44 out of 50 operators, i.e. more than 80%, are temporary residents. Representing 51.2% of the 86 temporary-resident operators of tourist souvenir shops, this figure indicates that the flood of products without local characteristics is associated to a considerable extent with the population influx. Thus it can be seen that, as things currently stand, the temporary residents are responsible for actively promoting a tourist industry based on commercialism, so much so that it owes its existence to the import of products from other localities that have nothing to do with the original local culture. As mentioned earlier, the system of family registration makes it difficult for temporary residents to take up permanent residence in a new locality. According to a government official in charge of such matters, this is an underlying reason why many operators tend to go to earn money in a prosperous tourist destination, only to move out and on to another locality when that destination ceases to prosper economically (based on interview with Lijiang County Bureau of Public Security, June 2000).

Local characteristics of products and attributes of shop operators

For Lijiang specialities, *Naxi*, with their permanent inhabitant status, could be expected to have advantages with regard to the supply, production and sale of such goods. Indeed more than half are handled by permanent inhabitants, predominantly *Naxi*. On the other hand, however, over 40% of such specialities are handled by temporary residents. A comparatively large number of such residents have begun to deal particularly in contemporary *Dongba* art objects, a recent development in

Table 13.5. Lijiang: types of tourist souvenir shops and shop operators as classified by product characteristics.

Categories of local characteristics of products	Categories of products	No. of shops	Categories of shop operators			Temporary residents' share (%)	Place of permanent-residential registration of temporary residents (number of shop operators)
			Permanent inhabitants	Temporary residents	Corporate entities		
Lijiang specialities	Contemporary Dongba art[a]	27	14	13	0	48.1	Within Yunnan(3), Guangxi(2), others(8)
	Herbal medicines, tea leaf, local liquor[b]	17	8	8	1	47.1	Within Yunnan(2), Sichuan(2), Anhui(2)
	Gold, silverwork[c]	5	3	2	0	40.0	Dali(1), Fujian(1), others(3)
	Local books[d]	4	4	0	0	0.0	
	Sub-total	53 (36.6%)					
Specialities of elsewhere	Jade and woodcarvings[e]	21	0	21	0	100.0	Fujian(14), Ruili(4), others(3)
	Jade[f]	15	2	13	0	86.7	Fujian(4), Henan(2), Myanmar(2)
	Specialities of other ethnic groups or regions[g]	7	1	6	0	85.7	Guizhou(2), others(4)
	Indigo-dyed goods[h]	5	2	3	0	60.0	Dali(2), Guangxi(1)
	Woodcarvings[i]	2	1	1	0	50.0	Fujian(1)
	Sub-total	50 (34.5%)					
Products deficient in local characteristics	General souvenir items[j]	28	16	11	1	39.3	Kunming(6), others(5)
	Ethnic-style clothes[k]	7	2	4	1	57.1	Henan(2), Dali(1), Heilongjiang(1)
	Antiques[l]	6	2	4	0	66.7	Heqing(3), Guangxi(1)
	Sub-total	41 (28.3%)					
Others	Ticket office[m]	1 (0.7%)	0	0	1	0.0	
Overall		145 (100.0%)	55	86	4	59.3	

[a] e.g. handicraft items drawn with traditional cultural motifs of the *Naxi*, such as *Dongba* pictographs, religious paintings (e.g. pottery, gourds, T-shirts).
[b] e.g. herbs and tea leaves produced in the alpine belt of Lijiang, liquor made from barley or sorghum.
[c] e.g. a traditional industry of Lijiang (e.g. necklaces, bracelets).
[d] e.g. books published in Lijiang, collections of photographs of Lijiang, postcards.
[e] e.g. shops that provide both jade and woodcarvings.
[f] e.g. shops that provide only jade, which is a speciality of Myanmar.
[g] e.g. handicrafts of minority races (e.g. *Miaos*, *Kazaks*, Tibetans).
[h] e.g. handicrafts of the *Bais*.
[i] e.g. shops that provide only woodcarvings, which are specialities of the southern regions of Yunnan Province.
[j] e.g. items that are available everywhere in Yunnan Province; mass-produced products, such as money pouches, dolls.
[k] e.g. items that are available everywhere in Yunnan Province; mass-produced imitations of native dress.
[l] e.g. old coins, old photos, Buddhist images.
[m] e.g. tickets for the cable car of Jade Dragon Snow Mountain.
Note: Source of speciality product or service ascertained from in-person interview with shop and a referenced source (Guo, 1999).

Fig. 13.9. Lijiang: an example of contemporary *Dongba* art objects.

handicraft items characterized by traditional cultural motifs of the *Naxi* (Fig. 13.9). Of the 27 operators here, almost half (13) are temporary residents (Table 13.5), of whom 12 are *Han*. Relative proportions of permanent inhabitants and temporary residents were *Naxi* 12 : *Han* 2; *Han* 12 : *Naxi* 1, respectively.

Contemporary *Dongba* art objects involve modern settings of *Dongba* pictographs, the *Naxi*'s unique pictographic form of writing, or of *Dongba* pictures, religious drawings and paintings used in *Dongba* religion, the *Naxi*'s traditional and peculiar faith. The artists bring an original resourcefulness to bear in endowing the surfaces of multifarious materials with pictures and colouring. Shops take the form of workshops that produce and sell the artefacts on site.

According to information revealed in an interview survey, Mr A, a *Naxi* craftsman, began creating in 1996 or thereabouts carved pictures featuring *Dongba* pictographs and *Dongba* pictures. Extremely elaborate and of great artistic value, his creations were thus expensive and have failed even now to establish a place for themselves as souvenir items. On the other hand, the people of the incoming *Han* population, despite having had no contact with the culture of the *Naxi*, have apparently begun to create some reasonably priced (RMB10–20 per item) products that, albeit crude, are engaging

as souvenir items for no other reason than that the *Han* excel at fashioning merchandise (information gained through multiple interviews with operators, May 2000). While this indicates that the incoming *Han* play a significant role in the processes involved in the commercialization of products and in present-day business management, it also suggests that the local community, which would normally be the perpetuator of its culture, has lacked the cumulative know-how to create a successful industry.

Turning to specialities from elsewhere, it was found that jade and woodcarvings occupy the greatest number of shops, both often being sold in the same store. Almost all of such stores are operated by temporary residents. Also evident is a concentration of operators from a particular locality, in that, of the 38 shops selling both jade and woodcarvings or one of the two, 50% (19 operators) hailed from Fujian Province. Many points remain unclear as to what lies behind these developments, and further investigations into this subject are needed.

Comparatively large numbers of people, both temporary residents and permanent inhabitants, engage in business involving general souvenir items. Such products can be seen everywhere in Yunnan Province, from which it can be inferred that extensive distribution networks make those items relatively easy to stock and that this is an easy type of business to take up for both permanent inhabitants and temporary residents alike.

Partly no doubt because few people are entering any business related to products other than those described above, no significant trend in operator attributes was observable in areas related to such other products.

Official response to goods lacking local characteristics

Anyone wishing to operate a store in Lijiang currently needs to apply for permission from both the Bureau of Commerce and Industry and the Bureau of Taxation and then be granted the respective permits: a 'Commerce and Industry Operator's Permit' and a 'Certificate of Tax Registration'. So far, however, there have been no restrictions based on the family register status of the operator or on the items for sale. As for the system governing the conservation of the old town, the principal restriction was on the external appearance of the building façade, while there were no rules governing how building interiors were to be used, or the nature of the products to be sold. However, amid the rising prosperity of the tourist industry and the attendant flood of specialities from elsewhere, both town and county governments began to recognize that such products were 'not in keeping with the characteristics of the old town and a negative influence on the promotion of tourism'. As a result, from December 1999, the authorities prohibited the opening of any further new shops engaging in the sale of jade and woodcarvings in the central section of the old town. At the same time, it was decided to start encouraging the sale of Lijiang's own distinctive handicrafts and to give preference to the sale of artefacts fashioned in silver and *Dongba* art objects when awarding permits to operate in the central section of the town. None the less, the authorities have gone no further than the preferential granting of operating permits and have adopted no policies to support the local people in developing industry, whether providing aid to business entrepreneurs or education and training in the skills needed to produce handicrafts (UNESCO Asia and Pacific Regional Bureau, 2000a; J.F. He, mayor of Dayan town, 2000, personal communication).

Discussion

This chapter has examined the nature of shops in the central section of the old town of Lijiang with a view to ascertaining the ways in which both the existing local community and temporary residents engage in the tourist industry. It has assessed business categories, operator attributes, forms of shop ownership, and descriptions of the goods on sale. The following three points emerge from a summary of the results of this analysis.

- The business category of tourism has been growing in the locality surveyed since about 1995. This increase is in large part due to the influx of an external population, with more than half of tourism being managed by temporary residents.
- Over 60% of these temporary residents pay high rents for premises in the private sector in which to run their shops. This indicates that the presence of a rental market, predominantly property in the private sector owned by the *Naxi*, has made it possible for temporary residents to enter the tourism industry.
- Approximately two-thirds of goods sold at souvenir shops are either devoid of or lack local characteristics. The greater part of the shop operators are temporary residents, who run more than 80% of the stores selling specialities from elsewhere. Although moves to create new products that bring to bear distinctive regional culture are also emerging, temporary residents play a major role in developing goods for the market and in business management. Accounting for approximately 40% of operators in the business category of tourism, the permanent inhabitants, the *Naxi*, on the other hand, while making their own significant contribution to the tourist industry, cannot, however, be described as displaying sufficient independent initiative in the creation of new industry owing to the absence of government policies for the support of local industry.

As noted above, Chinese people now enjoy a degree of freedom of movement, while the country is also seeing the development of a real-estate market. In such circumstances and according to market principles, there will naturally come to reside in the locality people capable of putting the land and buildings to effective use. As to the tourism business, it is easy to imagine that this will mean the influx of people that possess sufficient capital to open shops and sufficient know-how to raise profits. Further, given the lack of government policy for providing the local community with aid to develop its industry, temporary residents are likely to have an advantage in store management over existing inhabitants, in that the temporary residents have more financial capital and experience. Such a situation will see an increasingly greater proportion of temporary residents coming to operate tourism-related shops, while many former inhabitants will see it more to their advantage to move out of the old town and live off the rents accrued from leasing their former rooms to the temporary residents.

Temporary residents are applying business specialization and aggressively developing the tourism industry along commercial principles, while the *Naxi*, who are the successors to the original local culture, have proved incapable of bringing to bear sufficient independent initiative in the creation of the tourism industry. The result has been that the content of much of the tourism industry has little contextual bearing on the locality. This gives rise to the problem that tourism-related products do not successfully serve in communicating the allure of the locality, thus adversely affecting the process of host–guest exchange.

Conclusions and Implications

In the light of what has been described above, one can point to the following challenges that remain with regard to the existing tourism industry in this locality.

It will first be essential to include planning formulated in accordance with market principles in any policy for the preservation of the old town. It is necessary to foster the kind of local industry that will enable the local community to keep abreast of the high economic value and high rental charges of the protected buildings. If this is to be achieved, it will be necessary to take squarely into account the fact that, despite the opportunities of the local community to

commercialize its own distinctive culture, as exemplified by the case of contemporary *Dongba* art, the lack of specific forms of support has failed to create an internally generated industry. The task ahead thus needs to include policies for the promotion and aid of local industry, which will specifically involve:

- training to assist putting handicrafts on a commercial footing; and
- the provision of systems to aid entrepreneurs.

Once such policies have stimulated organizations and networks, and thereby invigorated attempts to create industry based on the local context, tourism-related activities can come to play a part in conveying local appeal to visitors and enabling the original culture of the heritage site to be perpetuated.

A further important issue to address is how temporary residents can best use their capital and know-how to engage in tourism-related activities as hosts best qualified to convey local appeal.

This chapter has taken as its subject the central section of the old town of Lijiang World Heritage Site, which is quite limited in scope, while the indicators used as the basis for its analysis have also been quite limited. It is thus necessary to regard the findings of this study as of qualified usefulness. What they do strongly suggest, however, is that the current architectural regulations alone, solely focusing as they do on the physical aspects of the preservation of heritage, are insufficient if the old town is to be adequately preserved. Instead, there would seem to be a need to conceive of protecting the heritage site as an entire entity that includes the structure of the local community. Such a perspective, moreover, could be applied not only to World Heritage city sites, but also extensively with respect to other historical cities and cities attempting to promote tourism.

Acknowledgements

The author wishes to thank Dr Tian-Xin Zhang, Beijing University, China, and Ai-Jun He, University of Tokyo, Japan, for their encouragement and help throughout the research upon which this chapter is based.

References

Drdácký, M.F. (2002) Impact and Risks of Tourism in Cultural Heritage Environment. Proceedings of the *Scientific Conference: Preservation, Development and Monitoring of Historic Cities in 21st Century*. ICOMOS, Athens.

Goullart, P. (1955) *Forgotten Kingdom*. John Murray, London.

Government of Yunnan Province (1997) *Conservation Detail Plan for Lijiang Old Town*. Government of Yunnan Province, Kunming.

Guo, D. (ed.) (1999) *General Survey of the Naxi Culture*. The Nationalities Publishing House of Yunnan, Kunming.

Guo, J., Duan, Y. and Yang, F. (1999) *Introduction to Minority Tribes in Yunnan*. Yunnan People's Publishing House, Kunming.

Lijiang Prefecture Administration Office (1997) *Lijiang Almanac 1997*. Yunnan Science and Technology Press, Kunming.

Martin, A.L. (2000) Making tourism sustainable. *Source* 120, 10–11.

National Bureau of Statistics of China (1999) *China Statistical Year Book 1999*. China Statistical Information and Consultancy Service Centre, Beijing.

Smith, V. (1989) *Hosts and Guests: the Anthropology of Tourism*. University of Pennsylvania Press, Philadelphia, Pennsylvania.

UNESCO Asia and Pacific Regional Bureau (2000a) Action Plan for Lijiang. Report of UNESCO conference/workshop, *Enhancement of Stakeholder Cooperation in Tourism Development and Heritage Preservation in Asia and Pacific*, Bhaktapur, Nepal, 8–16 April. UNESCO Asia and Pacific Regional Bureau, Bangkok.

UNESCO Asia and Pacific Regional Bureau (2000b) A Heritage Protection and Tourism Development: Case Study of Lijiang Ancient Town China. Paper for UNESCO conference/workshop, *Enhancement of Stakeholder Cooperation in Tourism Development and Heritage Preservation in Asia and Pacific*, Bhaktapur, Nepal, 8–16 April. UNESCO Asia and Pacific Regional Bureau, Bangkok.

World Bank (1998) *Culture and Development at the Millennium: the Challenge and the Response*. World Bank, Washington, DC.

Xu, J. (ed.) (2000) *Yunnan 'Tu-Dian' Illustrated Book: Lijiang*. Yunnan People's Publishing House, Kunming.

14 Tourism administration and regional integration under transition: policy and practice in South Africa

Jenny Briedenhann and Steve Butts

Introduction

South Africa has emerged from a long political history in which the majority of the population was debarred from tourism. The 1994 transformation of government set in motion a series of political changes (Giliomee and Schlemmer, 1993). However, transition to a political democracy has not led to an equivalent economic parity. In a country of high population growth, a weakening currency and rampant unemployment, tourism has been envisaged as a way to stimulate employment generation and a redistribution of economic wealth throughout the country's rural areas.

The transition period for tourism was marked initially by the Director of Tourism Promotion for South African Tourism proposing, in 1994, a 'new tourism' for the country (Heath, 1994; Brennan and Allen, 2001). This was followed by the formulation of a National Tourism White Paper, *Development and Promotion of Tourism in South Africa*, published in June 1996 by the Department of Environmental Affairs and Tourism (DEAT, 1996). This was followed by the strategy document *Tourism in Gear* (DEAT, 1998). These have been complemented by a plethora of provincial tourism white papers and legislation. However, lack of coordination between national and provincial government departments, pertaining both to tourism and its

related sectors, is rife. This was aggravated by the establishment of a National Tourism Organization (NTO), South African Tourism (formerly known as SATOUR), and nine Provincial Tourism agencies, leading to duplication, lack of synergy and waste of scarce human and financial resources. Competition between provinces further fragmented national efforts, bringing in its wake a confusion of destination images in the minds of potential consumers. In some provinces, regional district councils are heavily involved in tourism development and promotion. In other areas such councils refute any responsibility for tourism. Although the responsibility for tourism rests at local government level, most councils function in a vacuum with regard to tourism policy, planning and development, and have a critical lack of understanding of the industry as well as limited capacity to implement policy. However, this is not unique to South Africa. As Hall and Jenkins (1995: 29) point out, in many countries the policies and programmes of government are poorly integrated, on occasion contradictory and frequently implemented with minimal consultation between levels (see also Chapter 10).

International tourism to South Africa, after phenomenal initial growth rates of up to 66% in the first half of the 1990s (TTI, 1999), had slowed down by 1998 to just 3.5% growth. In 1998 the travel and tourism industry directly accounted for 2.4% of total employment and

2.6% of total GDP. With indirect impacts, it contributed 7% of total employment, 13.2% of total exports, 11.4% of total investment and 8.2% of total GDP (World Travel and Tourism Council, 1998: 10). Despite recently increasing tourist numbers (Table 14.1), the avidly anticipated tourism boom has not materialized. Desperately needed job creation and poverty alleviation is not forthcoming. The tourism miracle, touted by government as a panacea for the country's socio-economic ills, is losing its lustre. Yet the country encompasses a wide range of natural and cultural attractions (Figs 14.1 and 14.2).

This chapter seeks to understand the causes and consequences of tourism policy, and its implementation in South Africa, simultaneously examining policy objectives and integration processes at national, provincial, regional and local government level. The structure of tourism policy following the transition period to a political democracy is analysed, institutional arrangements identified, and success factors and constraints interrogated (Akehurst *et al.*, 1994: 11).

Tourism Policy Formulation

Tourism policy is influenced by the economic, social and cultural circumstances of a country

Table 14.1. South Africa: travel and tourism economy aggregates.

Year	Employment (in 000's)	Income (in Rand mn)
1998	1079.60	51.96
1999	1060.80	54.51
2000	1089.30	60.46
2001	1118.55	66.64
2002e	1147.98	72.46
2002p	1507.04	176.06

e, estimate; p, projected.
Source: The Department of Environmental Affairs and Tourism, Tourism Research and Statistics Division, Pretoria.

together with prevailing structures of government and features of the political system. Williams and Shaw (1988: 230) point out that different levels of government tend to have diverse tourism objectives and that the aims of local authorities may bifurcate from those of national government. This situation particularly prevails in federal systems akin to that of South Africa, composed of national, provincial, regional and local levels of government. Under such circumstances some groups exert greater influence if policy is formulated at national level whilst others may benefit more if policy is made at state or provincial level (Anderson, 1984: 18).

Fig. 14.1. South Africa: within the Drakensburg Mountains.

Fig. 14.2. South Africa: Cape Town from Table Mountain.

Pearce (1992: 4) states that tourism public policies are entangled in a dynamic, ongoing process during which it is evident that governments struggle to comprehend the tourism industry and their requisite role. This confusion manifests itself in the South African context by the fact that tourism has, over time, been relegated to a secondary function in departments with responsibilities as diverse as fisheries, forestry and water affairs. Currently, resources allocated to tourism at national government level come a poor second to those allocated to most other sectors.

De Kadt (1979: 33) suggests that the type and extent of tourism policy and planning are influenced most by a nation's ideology and its overall social, political and economic goals. Gunn (1988: 62–63) states 'policies and practices for tourism will follow the overall policies and practices of the nation as a whole. This is reflected in the relative roles of government and private enterprise, how profits are divided, the sectors most likely to benefit, domestic versus foreign travel influence, and relative dependence on tourism.' Government may prescribe policy pertaining to national tourism marketing and promotion initiatives, in some instances choosing to focus on domestic tourism development in preference to international tourism. In developing countries, domestic tourism may

be used as a catalyst for redistribution of economic benefits amongst income groups, from urban to rural and developed to undeveloped areas (Archer, 1977: 14). South Africa has a huge, untapped emerging domestic market, yet little is understood of the needs and preferences of this sector, and efforts to research the market are minimal (Bennett, 1995: Bartis, 1998).

Inskeep (1991: 29) contends that tourism development policy should be realistic and implementable, and that implementation techniques should be considered, and identified, throughout the process of formulation. In South Africa, substantial financial resources have been expended on formulation of tourism policies and strategies, few of which are implemented. This is primarily attributable to lack of capacity amongst delegated officials, a dearth of understanding of the tourism industry, lack of cohesion and integration with other economic sectors and severe budgetary constraints. In the South African context it has become obvious that, whilst policy statements in the National Tourism White Paper are appropriate, implementation and delivery are not forthcoming, leaving expectations unrealized and leading to frustration and disillusionment.

Middleton and Hawkins (1998) contend that policies are best understood when targeted at local, rather than national or regional level, and have specific significance for the economic and employment needs of developing countries with otherwise limited resources to sustain the demands of rapidly increasing populations. A critical restriction on tourism development is the lack of knowledge and understanding of tourism in the public sector, particularly at local level. In South Africa this presents a cardinal constraint, since expectations of tourism as a potential source for the alleviation of poverty and socio-economic ills run dangerously high, with little understanding of the implications of potentially negative impacts which tourism brings in its wake (Middleton and Hawkins, 1998: 105). Furthermore, in South Africa research into resource allocation and available capacity is limited, and official statistics proffered are highly questionable.

A coherent, realistic national tourism policy establishes the basis for developing and maintaining tourism and is an essential element of tourism planning and decision making. Inskeep (1991: 432) believes tourism policy should first be formulated on a preliminary basis and finalized only after it has been tested for suitability in achieving objectives and implementation feasibility.

Methodology

Policy integration and synthesis, together with clear role clarification and lack of capacity, are the most critical issues facing the development and management of tourism in South Africa today. In the absence of a national tourism growth strategy, vast sums are expended on payments to consultants, contracted to prepare provincial, regional and local tourism policies, strategies and growth plans. Fragmentation is diluting benefits accruing from tourism and expectations raised by the National Tourism White Paper are not being realized.

To understand better the prevailing status of tourism policy and its implementation at all levels of government, a mixed method approach was employed. Fieldwork, initially conducted in 1999, aimed to investigate the current status of tourism policy and implementation, and to formulate recommendations for future policy development and implementation

directives. Triangulation and testing of outcomes with a variety of stakeholders served to ensure credibility of data.

In seeking to gain understanding, and to identify the important dimensions of the problem, the principal researcher used practical experience and intimate knowledge of the issues under interrogation, together with authoritative status, as criteria in the selection of participants. Chosen candidates were deemed proficient enough to probe tourism policy issues and render worthwhile data.

Written policy documents were consulted and analysed. At the same time, questionnaires were issued to a purposive sample of 18 senior provincial tourism public sector role-players. Participants were drawn from both government departments and public sector tourism agencies. The questionnaire consisted of 14 questions. Questions relating to policy objectives, structures, roles and responsibilities were comprehensively answered. Those relating to the process of budget formulation and allocation generally evinced total ignorance both of government procedure and allocation of funding.

Open-ended interviews were conducted with 11 deliberately selected informants drawn from five of South Africa's nine provinces. Selection was based on expertise – all held positions of authority, were well informed and privy to the most relevant and up-to-date information, with well-defined opinions as to the interpretation, applicability, success and constraints of current tourism policy as related to their particular situation. Since all interviewees were well known to the principal researcher on a professional level, participants were generally prepared to speak openly with regard to problems and frustrations encountered.

Interviewees included: provincial heads of departments; chief executive officers of provincial tourism agencies; chief executive officers of regional and local councils; executive officers of national tourism organizations, and consultants directly involved in facilitating the formulation of tourism white papers and the identification and planning of tourism development projects. Gender representation was poor, with only two female respondents to questionnaires and one female interviewee. This is indicative of South African tourism, where the majority of employees are female, yet upper echelon positions are predominantly held by males. Every attempt was made to include representatives of all ethnic groups, and just over half of respondents were non-white.

The reluctance of officials from the outset of the research process to become involved in a study with potentially political overtones quickly became apparent. Whilst reservations were not openly expressed, severe difficulty was experienced in persuading parties to complete questionnaires. Three provincial departmental officials, despite numerous requests, declined to participate in the survey. It was deemed that much of this reluctance was based on the fear that survey results could reflect negatively on participants themselves.

The 11 interviewees indicated willingness to discuss issues viewed as consequential to the success of tourism in South Africa. In a few instances participants appeared to relish the discussion as an opportunity to vent frustrations and air views previously undeclared. One participant, initially seeking assurances of anonymity, later professed relief at being able to speak freely. A few individuals, with long histories of public sector involvement, expressed themselves with caution.

Findings

Many officials, both political and administrative, now entrusted with responsibility for tourism, have been left reeling in an effort to come to grips with the complexities of a demanding industry operating in a competitive, dynamic, global environment. Reluctance by some officials to participate in analytical interrogation of issues pertaining to tourism stems from this feeling of inadequacy. In this section key findings are presented in relation to tourism policy and objectives.

Table 14.2 addresses the publication of provincial tourism policy and the promulgation of legislation. Of the nine provinces only one indicated that a provincial tourism policy had not been formulated. Four provinces indicated that they had not yet promulgated provincial tourism legislation.

Provinces were asked to rank a series of 16 policy objectives on a scale of 1–10 (1 = low and 10 = high), according to the importance accorded them in that particular province (see Table 14.3). This interrogation is based upon a survey of successful tourism policies in 12 member countries of the European Union (Akehurst *et al.*, 1994: 16). The European study found that policy objectives tend to

Table 14.2. South Africa: provincial tourism policy and legislation.

Name of province	Tourism White Paper	Date	Tourism legislation	Date
Western Cape	Yes[a]	1999	Yes	1997
Eastern Cape	Yes[b]	Undated	No	na
Northern Cape	Yes	1998	Yes	1999
Northern Province	No	na	No	na
Mpumalanga	Yes	1999	Yes	2003
North West Province	Yes[c]	1998	No	na
KwaZulu-Natal	Yes[d]	1996	Yes	1996
Free State	Yes	1997	Yes	1998
Gauteng	Yes	1997	Yes	1999

[a] The Western Cape tourism Green Paper was launched in September 1999. This was published as a provincial tourism White Paper in 2000.
[b] The Eastern Cape document is published as the third draft of the provincial tourism Green Paper.
[c] Documentation for the North West Province was published as *A Tourism Masterplan for the NorthWest Province of South Africa*. A section of the masterplan dealt specifically with the policy framework.
[d] KwaZulu-Natal commented that although they had a provincial tourism policy paper, it seemed not to have been finalized into a White Paper. A strategy which focused on key areas followed.
na, not available.

Table 14.3. South Africa: ranking of tourism policy objectives.

Policy objective	Provincial department	Provincial tourism agency	Average
Creation of employment	9.5	9.0	9.2
Integration of previously disadvantaged communities	9.3	9.2	9.2
Redistribution of tourism to rural and marginalized areas	9.3	9.0	9.1
Increase numbers of domestic tourists	8.5	9.3	8.9
Increase international tourist expenditure	8.7	9.1	8.9
Increase domestic tourist expenditure	8.7	9.0	8.9
Increase numbers of international tourists	8.3	9.1	8.7
Encouragement of joint venture and empowerment initiatives	9.0	7.9	8.45
Diversification of the tourism product	8.5	7.7	8.1
Training and professionalism	9.0	7.1	8.05
Improvement of product and service quality	8.8	7.2	8.0
Increase industry size	8.3	7.2	7.7
Protection of the environment	8.5	7.0	7.75
Provide expertise to the industry	7.5	5.9	6.7
Reduction of seasonality	6.7	6.4	6.5
Regulation of the industry	7.7	5.3	6.5

Ranked on a scale of 1–10: 1 = low level of importance and 10 = high level of importance.

concentrate on attracting more high-spending tourists, improvements in product quality and a reduction in seasonality. By contrast, in South Africa, prime objectives are the creation of employment, integration of previously disadvantaged communities into mainstream tourism, and redistribution of tourism to rural and previously marginalized areas. Achieving a reduction in seasonality was ranked as one of the lowest priorities.

Averages for North West Province, KwaZulu-Natal and Free State are based solely on ratings of the provincial tourism agency. Officials of the relevant government departments declined to participate. Nevertheless, overall the National Tourism White Paper, to which provincial policies are reasonably well aligned, is considered a good document. Participants in the survey concluded that the policy is not at fault. It is the implementation strategies which are flawed. What need to be worked out are the 'hows' of delivering. Unfortunately implementation relies heavily on levels of capacity and high levels of management in order to succeed. This capacity is generally not available amongst those who should be the implementing agents – the public sector tourism bodies, both government and its agencies.

Provinces were asked to rate the efficacy of their provincial tourism policy with regard to

implementation and techniques for monitoring and measuring policy impacts (see Table 14.4). Whilst there are notable differences in provinces' evaluation of the efficacy of their respective tourism policy, variances of opinion between government departments and tourism agencies are not marked. As regards monitoring of policy and measuring of tourism policy impacts, opinions differed vastly.

As indicated in Table 14.5, provinces were asked to ascertain the level of integration of provincial and national tourism policy. Whilst policies appear reasonably well aligned, the authors' belief that there is little coordination of either resources or effort, and minimal integration of programmes (Hall and Jenkins, 1995: 29) is substantiated by participants in the survey identifying issues regarded as primary constraints to implementation. While integration of national and provincial policy statements is considered satisfactory, the mapping of roles and responsibilities in order to implement such policy is viewed as deficient. This means that there is no clear delineation of roles and responsibilities. This leads to frequent confrontation between role players at national and provincial level and between government departments and their public sector agencies, which represents duplication of effort, confusion amongst stakeholders and wasting of scarce resources.

Table 14.4. South Africa: rating of tourism policy efficacy.

Name of province	Provincial tourism department		Provincial tourism agency	
	A	B	A	B
Western Cape	a	10	a	7
Eastern Cape	4	3	4	4
Northern Cape	6	4	8	10
Northern Province	b	b	b	7
Mpumalanga	7	6	8	1
North West Province	c		5	5
KwaZulu-Natal	c		5	5
Free State	c		6	4
Gauteng	7	7	5	2
Average	6	6	5.86	5

Ranked on a scale of 1–10: 1 = low level of efficacy and 10 = high level of efficacy.
A, implementation efficacy; B, techniques for monitoring and measuring policy impacts.
[a] Western Cape was of the opinion that, since their policy had only recently been formulated, it was too early to pass judgement.
[b] Northern Province did not have its own provincial tourism policy. The Provincial Tourism Agency maintained that techniques were in place to monitor the efficacy of national tourism policy as applied provincially.
[c] Government departments of North West Province, KwaZulu-Natal and Free State failed to respond.

Provinces were asked to rank the involvement of regional district councils in tourism in their respective provinces with regard to implementation and budget allocation (see Tables 14.6 and 14.7). Whilst Western Cape Province, North West Province and KwaZulu-Natal appear to have notable success in integrating regional district and metropolitan councils into both planning and implementation of programmes, in other provinces, contribution is poor.

Table 14.5. South Africa: integration of national and provincial tourism policies.

Name of province	Provincial tourism department	Provincial tourism agency
Mpumalanga	10	10
KwaZulu-Natal	c	10
North West Province	c	9
Northern Cape	9	8
Western Cape	8	8
Free State	c	8
Northern Province	10[b]	6[b]
Eastern Cape	6	a
Gauteng	7	5
Average	8.3	8

Integration ranked on a scale of 1–10: 1 = poorly integrated and 10 = well integrated.
[a] Question not completed.
[b] Northern Province did not have its own provincial tourism policy. Whilst the Department opined that the National Tourism White Paper was implicitly followed, the Provincial Tourism Agency disagreed.
[c] Government departments in North West Province, KwaZulu-Natal and Free State failed to respond.

The Regional Services Council Act 109 of 1985 purports to have a positive effect on tourism development providing resources for both marketing and community participation. Metropolitan councils, in particular, allocate budgets, frequently in excess of those awarded to provincial tourism agencies, to development and marketing. The point was made, however, that even in provinces supported by regional district councils, there are imbalances between cities and the provincial hinterlands, where councils are unable to provide effective funding.

In one province, councils have a marketing organization that works independently of the provincial agency. Councils accuse provincial agencies of unwillingness to acknowledge their expertise. Conversely, agencies accuse councils of working in isolation and state that there is no structured synergy in terms of campaigns and programmes.

Regional district councils (RDCs) have a vital role to play in South African tourism since they have strong links with people in the industry locally, and many have substantial budgets with which to implement projects. Areas in which RDCs are active have shown the greatest development delivery. Only KwaZulu-Natal and the Western Cape indicate that regional councils play a really constructive role in tourism in their provinces. It is in these two provinces where most developments have taken place that include rural communities and people from the previously disadvantaged sectors.

Respondents were also asked to rank the level of involvement of local government in tourism in their province (see Table 14.8). Local authority participation is perceived to be low in most provinces. While it is widely accepted that it is at local government level that delivery should take place, comments abound of the inability of institutions at this level to undertake their responsibilities (Middleton and Hawkins, 1998: 81–84). The consensus is that if local government cannot fulfil its roles it hinders everyone from delivering. Respondents emphasized local authorities' lack of understanding of tourism, inability to see opportunities, and limited knowledge of land use planning. To exacerbate matters, most local authorities are in dire financial straits. The lack of synergy, levels of capacity and tourism policy approach by local government in South Africa, is of fundamental concern when viewed in the context of the belief that 'the world's best hope of securing sustainability in travel and tourism lies not with national authorities, but with the competence and authority vested in local government responsible for specific tourism destinations, working in partnership with private sector business' (Middleton and Hawkins, 1998: 39).

There is no framework elucidated in the White Paper within which local tourism organizations should work. In areas where local government is not able, or refuses to assist in the establishment of tourism information centres, these are wont to collapse. The lack of marketing synergy again manifests itself at local level where institutions market their area in isolation to either the region or the province.

Only one province follows the lead of national government by allocating responsibility for tourism to the Department of Environmental Affairs and Tourism. In each province a government-funded provincial tourism agency has been established. Respondents were given a list of 11 tourism functions, and asked to indicate whether these are performed by the provincial government department or provincial tourism agency. Table 14.9 demonstrates that the number of functions either allocated to, or claimed by, both provincial tourism bodies is indicative of the confusion, duplication and waste of resources which occur at this level.

Table 14.6. South Africa: regional district council involvement in tourism viewed from provincial department perspective.

Name of province	Involvement in implementation			Budgetary contribution		
	A	B	C	A	B	C
Western Cape	4	4	4	0	0	0
Eastern Cape	5	7	5	3	6	5
Northern Cape	6	6	5	6	6	4
Northern Province	1	1	1	0	0	0
Mpumalanga	1	1	1	1	1	1
North West Province		a			a	
KwaZulu-Natal		a			a	
Free State		a			a	
Gauteng	10	10	10	1	1	1
Average	4.5	4.8	4.3	2.7	3.5	2.7

Regional district council involvement in tourism ranked on a scale of 1–10: 1 = no involvement and 10 = very involved.
A, tourism development; B, tourism marketing; C, tourism planning and management.
[a] Government departments of North West Province, KwaZulu-Natal and Free State failed to respond.

Table 14.7. South Africa: regional district council involvement in tourism viewed from a provincial tourism agency perspective.

Name of province	Regional district council involvement in implementation			Regional district council budgetary allocation to tourism		
	A	B	C	A	B	C
KwaZulu-Natal	10	10	10	10	10	10
Western Cape	8	7	5	8	7	5
North West Province	6	8	8	2	7	8
Eastern Cape	6	4	4	3	3	3
Northern Cape	3	5	5	2	6	1
Northern Province	2	2	2	2	2	2
Mpumalanga	4	4	5	1	1	1
Free State	3	3	3	3	3	3
Gauteng	5	3	3	1	1	1
Average	5.22	5.11	5	3.55	4.44	3.78

Regional district council involvement in tourism ranked on a scale of 1–10: 1 = no involvement and 10 = very involved.
A, tourism development; B, tourism marketing; C, tourism planning and management.

Table 14.8. South Africa: level of local government involvement in tourism.

Name of province	Provincial tourism department				Provincial tourism agency			
	A	B	C	D	A	B	C	D
Western Cape	8	8	9	10	6	8	7	9
Eastern Cape	6	5	7	7	3	5	3	5
Northern Cape	4	4	7	7	8	8	8	1
Northern Province	1	1	1	1	4	4	4	5
Mpumalanga	1	1	1	3	2	4	4	3
North West Province		a			7	7	7	6
KwaZulu-Natal		a			6	7	10	10
Free State		a			2	2	2	5
Gauteng	7	7	7	7	2	3	5	3
Average	4.5	4.3	5.3	5.8	4.44	5.33	5.55	5.22

Local government involvement in tourism ranked on a scale of 1–10: 1 = no involvement and 10 = very involved.
A, tourism planning; B, tourism development; C, tourism marketing; D, provision of tourism information centres.
[a] Provincial government departments of North West Province, KwaZulu-Natal and Free State failed to respond.

Tourism Promotion and the Private Sector

Although not recognized in policy documents, the role of the private sector was raised by interviewees as being of fundamental importance to the future of successful policy implementation. The importance of its role is negated by the fact that the sector is disparate, badly organized and confuses government by its own agendas on which there is often no consensus view (Inskeep, 1991: 413).

Significantly, private sector participants stated that focus on the basic business principles of supply and demand had been lost, with over-emphasis on creating supply and virtually none on creation of demand. Fostering market demand, increasing tourist volumes, escalating market profile, and ensuring quality standards are seen as catalysts for developing the industry and creating an environment in which investors are encouraged.

Table 14.9. South Africa: allocation of tourism functions at provincial level (as percentages).

Function	Government department	Provincial tourism agency	Allocated or claimed as a function by both bodies
Tourism marketing	–	100	–
Tourism development	11	67	22
Tourism investment promotion	11	56	33
Standards and service excellence	–	1	29
Tourism infrastructure development	33	22	45
Research	–	67	33
Tourism information management	–	89	11
Tourism awareness	25	50	25
Tourism safety and security	45	22	33
Tourism training	37.5	37.5	5
Tourism road signage and route development	25	50	25

Note: Percentages are based on the number of replies to the relevant questions. Some respondents failed to answer particular sections of the question.

The private sector customarily views government initiatives with scepticism. Institutional confusion, and the proliferation of structures with a dearth of tourism expertise and capacity, is particularly viewed with cynicism. Accusations abound of private sector companies influencing provincial market offerings to the detriment of other products. This creates great tension in the industry. Provincial agencies' inability to mobilize industry role-players, and perceptions of poor governance, do nothing to raise their image in the eyes of the private sector. Provincial rivalry, fragmented programmes, diluted resources, unilateral actions on the part of provinces, and a lack of strong provincial leadership for tourism all add to private sector antagonism. Since the private sector has had no faith in government institutions it has tended to act autonomously, resulting in a lack of coordination between public sector institutional actors and private sector providers.

Problem Areas and Success Stories

Functions and responsibilities of all tiers of government, and its agencies, are spelt out in national and most provincial tourism policy documents. Despite this, effective coordination between provincial departments, regional district councils and local government, and confusion with regard to role clarification between provincial and local government and tourism agencies are perceptible as major constraints

(Inskeep, 1991: 431). Also, since 1996 the national department has been constantly subject to political change. This has engendered a period lacking in both political and operational leadership. Poor understanding of the importance of tourism at national government level meant that national tourism departments were inadequately, or inappropriately, staffed and resourced to carry out their functions.

What became known as the 'SATOUR saga' has been the leading factor in the country's marketing malaise. Post-1995 SATOUR (now South African Tourism) was overstaffed, underfunded, and underskilled. The transformation process of the national tourism organization (SATOUR) into an effective marketing body took far longer, and was subject to many more external influences, than originally anticipated. Bureaucracy and lack of flexibility further inhibited the transition. Absence of role clarity and vying for power between the national tourism department and SATOUR was a cause for concern, with unnecessary duplication of actions, and mixed messages sent both to provinces and industry.

Operating in an environment of chaos, SATOUR had been unable to give leadership with regard to integrated complementary marketing plans, or in the field of hospitality accreditation standards. Provinces, in the interim, pursued their own courses, undertaking overseas marketing trips, duplicating effort, confusing the marketplace and wasting chronically scarce marketing resources. Whilst the NTO regarded provincial tourism agencies as recalcitrant children, the agencies viewed the NTO as a rather lame duck. At the same time the private sector viewed the NTO as 'not tourism competent and toothless', and believed the NTO should be privatized. In 1999 a public/private sector partnership was formed with a collection of levies for international tourism marketing and a national marketing committee that works closely with the NTO.

Problems experienced at national level are reflected in the provinces. Whilst the role of provincial tourism agencies is defined in most policy documents, confusion, infighting and a breakdown in communication exists between provincial departments and their agencies, with politics perceived to be playing a key role. The problem is exacerbated by the fact that tourism agencies generally view government departments as lacking in skills and capacity. Tourism agencies are, however, generally viewed with suspicion by politicians who regard them as costly, often corrupt institutions, unwilling to be accountable to government. Paucity of budget, and inability of provinces to secure greater funding for tourism, renders both departments and agencies unable to deliver on most responsibilities. Lack of synergy, and dearth of communication between tourism bodies and other departments prevents integrated inter-departmental involvement in tourism development. The allocation of tourism as a concurrent responsibility to provinces has led to provincial rivalry, which has on occasion manifested itself in the international marketplace. Destructive competition abounds between urban and rural areas, metropolitan and local councils, town centres and those settlements which had been designated as black townships under apartheid.

Informants from individual provinces were asked to highlight areas of tourism policy in which difficulties in implementation and delivery were experienced, or conversely where notable success had occurred. Recurring problems identified fall into five major categories:

- inability to integrate the previously disadvantaged community and marginalized rural areas, into the tourism industry;
- lack of financial resources to achieve policy objectives;

- lack of integration and synergy between all tiers of government;
- inter-governmental, inter-organizational and intra-organizational relationships; and
- lack of capacity and skills at all levels.

Success stories claimed by various provinces include:

- improved partnerships, joint ventures and the implementation of well-developed and visible strategic marketing alliances and programmes between tourism agencies and the private sector;
- successes in community tourism development with an increasing number of previously disadvantaged communities involved in tourism activities;
- significant interest being created amongst new entrants to the tourism industry by means of tourism awareness programmes;
- the process of securing participation, consensus, support and mutual trust from stakeholders is growing, albeit slowly;
- the development of strategies and provincial tourism growth plans which give direction for future growth and development of tourism;
- the identification and packaging of investment projects; and
- strategies being implemented to ensure the establishment of integrated tourism information networks and the development of tourism route roads aimed at luring tourists off main routes and into areas previously neglected.

The real strength and value of the claimed successes are, however, debatable and require further interrogation in the light of the fact that several issues raised are also indicated as problem areas.

Participants were generally of the opinion that policy failure and lack of demonstrable delivery are not as a result of policy deficiency, but rather a consequence of factors such as role uncertainty, deficient capacity and inadequate funding. Similar patterns of concern manifested themselves in constraints identified in questionnaires and policy documents, and are compatible with issues recorded in the literature.

Recommendations

Policy is not a self-evident, independent behaviour fact. Policy acquires meaning because an observer perceives and interprets a course of action amid the confusions of a complex world (Heclo, 1974: 4). There have been more than enough White Papers, reports and 'talk-shops' about tourism. What is needed is strong leadership, and decision making that is strategic, practical and sensitive to the problems of developing tourism in a newly evolved democracy – not least of the problems is the fostering of unrealistic expectations of the benefits to be derived from tourism. As one of our informants stated,

> we make a mistake envisaging what a project could be 10 years hence, but poor rural people are going to have a faster expectation. So, lighting a fire in somebody, you've got to be very careful that the fire doesn't burn you and them at the same time. So it's a question of igniting a little glowing ember rather than this fierce fire.

Recommendations deemed appropriate and practical in alleviating critical constraints within the South African tourism development process include the following.

- Tourism should be acknowledged and managed as a priority at government level. A ministry dedicated to tourism, with a strong committed minister, can make a key contribution (Elliott, 1997: 54). In 1999, a new tourism minister was appointed who carried industry credibility and managed to give tourism a much higher profile at government level. For example, under his auspices transnational 'peace parks' have been promoted, with the South Africa/Botswana Kalahari National Park being the first to be established. He has also supported the commercialization of national parks. Non-core services have been outsourced and private developers have been given 20-year concessions in national parks throughout the country to build, own and operate eco-tourism facilities. Cutting through years of political in-fighting, he appointed a new National Grading Council. As a result, a quality grading and classification scheme for accommodation is operating, with the intention of diffusing good practice to other services.

- A clear, coherent central government strategy for developing tourism is required together with a common vision to which all role-players, both public and private sector, can subscribe.
- That the national minister for tourism should establish a ministerial tourism committee comprising colleagues drawn from departments with whom synergy is essential, to progress tourism development in an holistic and integrated manner. A similar forum should be constituted at provincial level.
- At provincial level a forum should be constituted in which the provincial tourism leaders meet on a biannual basis with chairpersons of regional district and metropolitan councils, and regional tourism associations, to facilitate tourism strategy for the province.
- The above forum should be informed by a coordinators' committee, comprising officials of provincial tourism agencies and regional district/metropolitan/local council tourism coordinators. It should meet regularly to discuss and agree on collaborative tourism programmes and strategies.
- Local government capacity building, in terms of tourism understanding, skills, and funding, is essential so that responsibilities can be carried out effectively as the implementers and coordinators of tourism policy at grass-roots level.
- Government intervention is required in order to overcome the duplication, wastage and confusion currently permeating public sector structures. Strong governmental co-ordinating mechanisms should be instituted.
- The issue of a cohesive tourism institutional framework is a challenge worldwide as tourism is characterized as being fragmented and complex. A strategic mapping exercise should be undertaken as a matter of urgency, to clearly delineate functions, responsibilities and accountability of all tiers of government and its tourism agencies.
- The issue of a national tourism tax, or levy, and the establishment of a tourism growth fund, should receive attention at the highest level in order to obviate the introduction of unilateral funding mechanisms by provinces. A predetermined portion of funds collected should be earmarked for projects undertaken at grass-roots level. Although an international

departure tax has been implemented, the money collected goes straight into the central exchequer with no guarantee that any will filter down either to lower tiers of government or to tourism, where it is desperately needed for development.

- Policy relating to investment incentives is required which accelerates development, increases the speed of financial returns, and shows positive discrimination favouring preferred sectors or regions.
- Policy is needed which pursues full community consultation, participation and management of cultural tourism resources and programmes. Inherent in such policy is the need for protection against erosion of authenticity, violation of personal privacy and misuse of intellectual property.
- Lack of capacity is cited as a primary constraint to policy implementation. Tourism is one sector where actual experience and a deep knowledge of the industry and sector are required. Transfer of skills, training and tourism-awareness education must be encouraged at all levels of both public and private sectors. Government should ensure access to training for public sector employees, and devise incentives aimed at rewarding members of the private sector willing to mentor current, and aspiring, entrants to the tourism industry.
- Mechanisms should be devised to determine the impact of tourism policy, and measures which should be taken if performance deviates significantly from what is desired (Hall and Jenkins, 1995: 91). Retrospective reviews should be conducted to assess if impacts generated by development projects correspond to planned expectations (OECD, 1981). Proper monitoring measures should be devised to establish effectiveness indicators for marketing campaigns and to assess the impact of marketing expenditures compared with other factors influencing holiday-taking decisions (Deegan and Dineen, 1997: 220).

Conclusions

Whilst tourism policy should achieve a balance between public interest and government princi-

ples, culminating in a drive for a more dynamic, efficient and effective industry, the ultimate test of successful tourism policy formulation lies in its implementation. Non-implementation means there has been a waste of resources spent in policy formulation, expectations have been raised, and not realized, and the standing of both public-sector managers and politicians has been damaged. Lack of implementation, and delivery at grass-roots level, recurrently emerge as the most prevalent issues in the tourism policy arena.

Overall it would appear that in the governance of South African tourism, monitoring and evaluation, at all levels, is extremely poor, with government lacking mechanisms to ensure that policy is implemented and objectives realized. The significance of monitoring and evaluating policy and its implementation is paramount, since it is the actual impact of tourism policy that is ultimately of consequence. Constant monitoring alerts both decision and policy makers to situations in which public officials carry out different activities from those envisaged, or where policies fail to reach intended clients (Hollick, 1993: 125; Hall and Jenkins, 1995: 82).

The vast majority of South Africa's visitors are attracted by natural, wildlife and scenic elements. Future funding for environmental preservation and management is none the less a cause for concern. Amongst the communities are bitter opponents of conservation who have been forced to relocate to inferior land in order that reserves can be secured for affluent tourists. Only demonstrable delivery can convince poverty-stricken communities that tourism can provide employment and bring financial benefits to affected residents. The inability to stimulate employment or integrate marginalized rural areas and the disadvantaged community into the tourism industry, either as tourists or business people, is viewed as a fundamental policy failure. Considered strategically unimportant, tourism has suffered from restricted budgetary allocations. Paucity of funds to implement policy and to achieve objectives is a primary impediment identified by all role players. There are too many organizations seeking funding and a growing impatience on the part of the private sector. Government and private sector are far apart in terms of attitude and anticipated outcomes. Expectations have been widely raised in communities by virtue of consultation and identification of potential community-based tourism projects. Non-delivery on these developments may have serious political consequences.

Formulation and adoption of appropriate tourism policy is the essential first step in tourism development. This task is now complete. There is urgent need for the drafting of a congruent, coherent national tourism strategy aimed at stimulating job creation, entrepreneurial development and environmentally friendly, sustainable, rural tourism. The mapping of a coherent organizational structure, with a clear delineation of functions and responsibilities, is imperative in order to ensure policy execution. No amount of funding will solve the current impasse unless government ensures that these measures are in place and continuously monitored.

The case of South African tourism is somewhat different from that in most 'developing' countries. For example, high quality road and air infrastructure exists in many parts of the country. However, in marginalized rural areas where development is critically needed, access is problematic. Also, unlike that in most developing countries, the majority of the tourism superstructure is owned by South Africans as opposed to foreign investors. This is at least partly a result of the fact that until 1994 South Africa carried a stigma for potential investors. The challenge for South Africa now is to sustain the quality of existing attractions and infrastructure while at the same time creating opportunity and building capacity which will allow previously disadvantaged communities to share equitably in the benefits of tourism.

References

Akehurst, G., Bland, N. and Michael, N. (1994) Successful tourism policies in the European Union. *Journal of Vacation Marketing* 1(1), 11–27.

Anderson, J.E. (1984) *Public Policy Making*, 3rd edn. CBS College Publishing, New York.

Archer, B. (1977) Sustainable tourism: an economist's viewpoint. In: Briguglio, B., Archer, B., Jafari, J. and Wall, G. (eds) *Sustainable Tourism in Islands and Small States: Issues and Policies*. Pinter, London, pp. 11–22.

Bartis, H. (1998) A national black heritage trail in the Eastern Cape: is it an option? In: Hall, D. and O'Hanlon, L. (eds) *Rural Tourism Management: Sustainable Options*. SAC, Auchincruive, UK, pp. 17–28.

Bennett, J. (ed.) (1995) *Managing Tourism Services*. Van Schaik, Pretoria.

Brennan, F. and Allen, G. (2001) Community-based ecotourism, social exclusion and the changing political economy of KwaZulu-Natal, South Africa. In: Harrison, D. (ed.) *Tourism and the Less Developed World: Issues and Case Studies*. CAB International, Wallingford, UK, pp. 203–221.

DEAT (Department of Environmental Affairs and Tourism) (1996) *Development and Promotion of Tourism in South Africa*. Government Printer, Pretoria.

DEAT (Department of Environmental Affairs and Tourism) (1998) *Tourism in Gear: Tourism Development Strategy 1998–2000*. Government Printer, Pretoria.

Deegan, J. and Dineen, D.A. (1997) *Tourism Policy and Performance: The Irish Experience*. International Thomson Business Press, London.

De Kadt, E. (1979) *Tourism: Passport to Development?* Oxford University Press, New York.

Elliott, J. (1997) *Tourism: Politics and Public Sector Management*. Routledge, London.

Giliomee, H. and Schlemmer, H. (1993) *From Apartheid to Nation-building: Contemporary South African Debates*. Oxford University Press, Cape Town.

Gunn, C.A. (1988) *Tourism Planning*, 2nd edn. Taylor and Francis, New York.

Hall, C.M. and Jenkins, J.M. (1995) *Tourism and Public Policy*. Routledge, London.

Heath, E. (1994) Into the 'new tourism'. *SATOUR Quarterly Review* 3(1), 2–3.

Heclo, H. (1974) *Modern Social Politics in Britain and Sweden*. Yale University Press, New Haven, Connecticut.

Hollick, M. (1993) *An Introduction to Project Evaluation*. Longman Cheshire, Melbourne.

Inskeep, E. (1991) *Tourism Planning: An Integrated and Sustainable Development Approach*. John Wiley & Sons, New York.

Kruger-Cloete, E. (1995) Funding the development of tourism in South Africa. *Development Southern Africa* 12(5), 751–761.

Middleton, V.T.C. and Hawkins, R. (1998) *Sustainable Tourism. A Marketing Perspective.* Butterworth-Heinemann, Oxford.

OECD (Organization for Economic Cooperation and Development) (1981) *Case Studies of the Impact of Tourism on the Environment.* OECD, Paris.

Pearce, D.G. (1992) *Tourist Organisations.* Longman, Harlow, UK.

TTI (Travel and Tourism Intelligence) (1999) South Africa. *TTI Country Reports* 2, 79–100.

Williams, A.M. and Shaw, G. (eds) (1988) *Tourism and Economic Development: Western European Experiences.* Belhaven Press, London.

World Travel and Tourism Council (WTTC) (1998) *South Africa's Travel and Tourism: Economic Driver for the 21st Century.* WTTC, London.

15 Conclusions and future agenda

Derek Hall and Lesley Roberts

This volume has been concerned with the ways in which tourism intersects with contemporary processes of capitalist expansion as exemplified by countries and regions in 'transition' or 'transformation'. This has involved addressing tourism's roles and development within a relatively wide range of experiences:

- the adoption of ('Western') democratic political processes and market economies in some, but not all, former Soviet bloc societies;
- preparation for EU enlargement to incorporate some European post-communist societies;
- the adoption of market-related economic systems within persisting one-party communist political systems;
- post-apartheid South Africa;
- post-division Cyprus;
- Malta, where issues of EU accession represent a substantial political cleavage within society; and
- the understanding of and response to impacts of political instability and terrorism.

Throughout this volume, the interaction of cultural, economic and political factors in the processes of tourism development has been emphasized. As noted in Chapters 1 and 2, the term 'transition', appropriated by Western analysts for a specific post-communist agenda, appears less than satisfactory within the terms

of the issues and range of societies discussed in the foregoing chapters. The more open-ended notion of 'transformation' has been a preferred term for a number of reasons. First, 'transition' has several interpretations and has long been conceptualized in a number of contexts, such as the 'demographic transition'. Somewhat paradoxically, however, the post-1989 application and appropriation of 'transition' is both prescriptive in character and ideological in intent. Such prescription would exclude a number of societies featured in this volume:

- those – such as EU entrants Malta and Cyprus – which do not have state socialist histories;
- those post-communist states ineligible for EU or NATO membership, such as Kyrgyzstan;
- those which are neither 'post-communist' nor seeking EU accession, such as South Africa; and
- those continuing state socialist societies that have transformed their economic systems but not their political structures and processes, such as China.

Finally, in seeking to provide a coherent framework for the range of societies in this volume, it is argued that notions of 'transformation' can help better to acknowledge the social and cultural processes which both inform and are informed by economic and political change.

Tourism and Transformation: Issues and Common Themes

Chapter 1 argued that the wide set of social, cultural, economic, political and environmental change agents, of which international tourism is part, are largely driven by the global expansion and rejuvenation of capitalism. This process inherently generates patterns of uneven development and inequality, which tourism may act to ameliorate or exacerbate in its not necessarily compatible roles as regional (sustainable) development tool and profit-oriented business.

The issues, problems, paradoxes and opportunities presented in subsequent chapters were diverse and pursued from different perspectives and at a number of levels, yet they share perspectives on ways in which tourism interacts with and within processes of transformation. These can be subsumed within two over-arching sets of issues: those of governance, and those of scale and linkage.

Issues of governance

Two key themes relating to the governance of tourism emerged in the preceding chapters. First was the frequent inability of institutions to respond adequately to the opportunities of dynamism and the possible threats of uncertainty in transformation. This was exemplified in Chapters 4, 10 and 14, from such contrasting contexts as Poland, North Cyprus and South Africa, raising practical issues of human training and institutional capacity-building on the one hand, and less tangible matters concerning political will and perceptions of strategic necessity on the other.

Second, questions relating to both the nature of the interface between, and compatibility of, political imperatives and socio-economic needs, were raised in Chapters 7, 9 and 12. These drew on the contrasting contexts of Serbia, Malta and China's Hunan province. The loosening role of the state, the need to meet quality requirements of supra-national organizations, and the nature and consequences of acceding to market forces, were seen as contexts reinforcing the reality that tourism development and impacts may both reflect and influence a broad and ever-changing matrix of relationships between political and socio-economic structures and processes.

Issues of scale and linkage

Four major themes of scale and linkage were brought to the fore. First, the ever-changing balance of endogamous and exogenous influences was well articulated in Chapters 6, 8, 9 and 13, drawing on evidence from Estonia, Kyrgyzstan, Malta and China. The tensions and realities of the often precarious equilibrium were expressed in such diverse areas as heritage interpretation, image construction, public transport provision and in the cultural and entrepreneurial roles of ethnic groups.

Second, and related to the first, the role and priorities of the local and the global, as represented in Chapters 5, 11 and 13, emphasized the contemporary significance and differential local impacts of such diverse global processes as foreign direct investment, terrorism and the conferment of World Heritage Site status.

Third, the potential tensions and synergies of partnerships and networks within and between private and public sectors – at local, regional and national levels – in shaping the path of tourism development, were reflected in Chapter 3, taking an overview of Central and Eastern Europe, and in the case of South Africa in Chapter 14.

Fourth, the interplay of tourism and non-tourism factors, which most chapters draw upon and highlight, reflects the persisting reality – perhaps particularly notable under conditions of transformation – that for analytical purposes it is often difficult, and may be impractical, to attempt to isolate the role of tourism from other dynamic social, economic and political processes in contributing to developmental outcomes.

Intersecting with these two sets of issues is the framework of pathway analysis, in which each transforming country is seen to have its own current and past determining factors influencing its trajectory. While not comprehensively applied, such a framework is implicit in many of the foregoing chapters that highlight specific national and regional characteristics particular to the circumstances of that country or localities within it.

Themes of transformation trajectory

Earlier chapters in the book focused on particular, if related themes, linking aspects of tourism development with economic and social pathways away from the ethos, processes and structures of state socialism. In the case of Chapter 3, through a case study from Romania, the importance of the recognition and mobilization of social capital was emphasized and pointed to an emerging literature in an area that is becoming increasingly recognized in economic development strategies. Qualities of trust, cooperation and cultural consensus have been credited with creating the conditions for embedded and sustainable economic development, in contrast to traditional performance factors such as costs and location. In the increasingly globalizing system, a parallel debate has concerned contradictions in promoting both economic cohesion and cultural diversity. Graham and Hart (1999) have argued that the European project in particular is unable to deliver these objectives simultaneously. This was echoed by Smith (2000) who raised questions about the EU's capacity to shape the 'New Europe', as noted in Chapter 2. Further, intersecting with issues of EU enlargement and the persistent nature of local cross-border activity, Grix and Knowles' (2002) recent discussion of the German–Polish Euroregion, Pro-Europa Viadrina (PEV), as a potential 'social capital maximizer' is instructive. They indicated how, despite expectations, the Euro-region fell short of playing a bridging role and fostering networks of reciprocal trust across the border.

For those states for whom the pathway from state socialism to EU membership has been completed in 15 years, particular challenges within trajectories of continuity and change have been highlighted. In Chapter 4, the focal issue for Poland was seen to be the question of state re-investment in tourism infrastructure and services. For Hungary, in Chapter 5, the complications of FDI and corporate ownership within the accommodation and catering sectors were highlighted. In both these national cases,

response in terms of tourism governance has often been ambiguous, and certainly influenced by the prescriptive pathway to EU accession. For Estonia, the ambivalent and changing role of cultural heritage before and after 1991, highlighted in Chapter 6, may again experience revision with EU enlargement embracing all three Baltic states.

The requirement of Serbia to re-establish a positive image and identity in Europe was emphasized in Chapter 7 through an evaluation of the use of imagery and the evolution of a national tourism policy based upon sustainability and explicit environmental friendliness – both natural and social. Unfortunately, portrayal of 'open doors' and a friendly folk ethic set within a pristine environment, as the smiling, welcoming image of Serbian tourism, has to contend with the legacy of more than a decade of negative media depiction in many parts of the world, regional conflict and ethnic polarization. Although largely a Western construction, Balkan 'otherness' appears to have some way to go before its pejorative connotations can be superseded by images of positive distinctiveness.

Such pathway constraints were echoed in Chapter 8 in the case of Kyrgyzstan. While lack of capital resources and political instability have constrained efforts to employ tourism as a means of positive image projection, the cultural paradox of notions of the 'other' have also pervaded the country's trajectory out of state socialism. The relationship between government policy, the construction of national identity and the development of tourism was articulated in the influences Kyrgyzstan's rich cultural heritage and ethnic diversity exert on image development. On the one hand, 'cultural otherness' is an important aspect of the spirit of place, which can reinforce the country's natural features and has the potential to evoke an emotional attachment. On the other hand, the awakening of clan, ethnic and religious identity, particularly marked in the south of the country, militates against the development of a national identity and a coherent tourism product and image. This conundrum helps to reinforce the sense of a vicious circle of peripherality out of which development policies will struggle to lift Kyrgyzstan.

Although superficially structurally better placed than most economies aspiring to full membership of the EU, Malta has faced a number of critical transformation issues on the pathway to accession. Chapter 9 indicated that with a limited natural resource base, central location in the Mediterranean and history of British colonial rule, the heritage of UK market-dominated mass tourism has constrained the country's diversification options. Increasing congestion within these small islands containing a limited number of centres reflects both the country's resource base and a post-colonial history of lax decision making exacerbated by a sharply divided political system which has perpetuated uncertainty about EU accession and the permanence of membership. As the Malta Tourism Authority has sought to establish an upgraded image and to pursue market segmentation to attract high-value niche tourists, so the combination of high domestic car ownership levels, a poor public transport infrastructure, an historic built environment, restricted resources and inadequate traffic management, have drawn tourism and transport development issues together into an intricate relationship subject to the further complications of EU accession requirements.

It might be said that until 1974 there were a number of parallels in the development of tourism in Malta and Cyprus, both in terms of their post-colonial market dependence and the nature of the tourism product available. However, as Chapter 10 indicated, following Turkish military intervention in the north in 1974, when the island was divided into separate geographical and political entities, Cyprus tourism, emblematic of wider social and economic dimensions, evolved as two different trajectories of development and growth. In marked contrast to the South, the trajectory in northern Cyprus (TRNC) has been strongly constrained by external forces of isolation and internal institutional shortcomings. While this combination of factors is unique to the Cyprus context, complicated further by the politics of pathways to EU accession, it echoes the paradoxes of competing endogamous and exogenous forces within issues of peripherality and in relation to the 'rules' of EU membership.

Taking a loose pathway approach to the evolving global role of tourism, alongside the oft overlooked dangers to tourists of disease and accidents (often self-induced), interna-

tional terrorism, both perceived and actual, may pose a lasting threat to global tourism's apparent current inexorable growth in both periphery and core destination regions. Chapter 11 argued that, more than ever before, there is now a primary need for addressing the interrelationships between, and issues related to, the sustainability of tourism growth and crisis recovery. If we conceive of international tourism as a (if not *the*) major symbol of the spread of global capitalism and 'Westernization', then it may be easy to see that, as an infrastructure-reliant, people-intensive face-to-face activity, tourism is both a major source of, and accessible target for, the hostility of various international terrorist and other political groups. In producing a theoretical and empirical framework to enable an evaluation of the relationship between political instability and tourism, Chapter 11 offers a tool which has been developed to assist the planning and management of tourism within such contexts of instability.

Chapters 12 and 13 illustrated some of the tensions and opportunities of employing tourism as a development vehicle for a state socialist society which has loosened economic control while maintaining a relatively tight centralized political command structure. Although focusing on different dimensions of tourism processes in specific regions of China, these two chapters highlighted some of the specific paradoxes and opportunities of tourism development set within pathway conditions of 'conservative transition'. Through the frameworks employed, they present opportunities for comparative analyses of tourism and governance both within the different components of China and between China and other societies (for example, see Rodriguez Pose and Gill, 2003). These chapters also suggest fruitful research avenues into factors influencing the level of priority given to tourism within Chinese development, particularly in comparison with, or perhaps complementary to, the shift of manufacturing FDI into the country from South-east Asia (see, for example, Felker, 2003).

South Africa has emerged from a long political history in which the majority of the population was excluded from tourism. The 1994 transformation of government set in motion a series of political changes; but political democracy did not lead to an equivalent economic and social transformation (an interesting corollary to the experiences of countries such as China and Vietnam). In a country of high population growth, a weakening currency and rampant unemployment, tourism has been envisaged as a way to stimulate employment generation and a redistribution of economic wealth throughout the country's impoverished rural areas. Within the context of these aspirations, Chapter 14 emphasized the fragmentary way in which newly 'liberated' decision makers at national, provincial, regional and local levels have generated plans, legislation and policies in the tourism field which have been insufficiently integrated or coordinated. Such a situation is not unique, and reflections on South Africa's transformation trajectory might suggest that the reported horizontal (spatial) and vertical (structural) incoherence is an inevitable short- to medium-term outcome of a system of governance still establishing itself. In the meantime, it poses a major constraint on Africa's largest tourism economy.

Conclusions: Issues of Transformation

Tourism and the European transition 'project'

This volume has been concerned with a wide range of 'transformational' societies, in terms of their prevailing ideologies, levels of economic development, geographical location, domestic and international aspirations and, not least, their tourism modes and profiles.

Acknowledging that it has not been an easy task to establish conceptual and methodological frameworks within which to accommodate all of these conditions, we return, in conclusion, to the specifically European post-communist transition 'project', not least because it has tended to dominate (at least European) thinking on 'transition' for the past decade and a half. This was earlier characterized as a prescriptive, ideologically informed Euro-Atlantic conception of 'transition' as a process of restructuring formerly communist political economies with the end goal of establishing economic, political and administrative norms which conform to the requirements for successful EU accession.

We now take the specific post-communist European 'transition' experience and place it within the simple generic tourism transition model suggested in Chapter 1, to highlight some of the implications for tourism of the goal of such transition, EU accession. Thus Fig. 15.1 characterizes tourism's transition within a three-aspect post-communist tourism sequence. First, it reiterates the societal context for post-communist tourism development indicated in Fig. 1.1. Second, through the tourism transition model's characterization of the sector's structure and spatial nature, it indicates some of the key challenges for post-communist societies' tourism development. Finally, the EU accession context, as an end point of post-communist transition, is addressed, raising implications for the tourism transition model.

Although such a focus serves explicitly to generate a formative analytical frame for development within the post-communist context, it also has the potential to be extended to other contexts of transformation considered in this volume. As such, Fig. 15.1 is not intended as a definitive model, but rather one that reflects some of the dynamics of contemporary transformations.

An agenda for further research

The scope and complexity of issues and contexts presented in this volume provide us with a potentially endless range of questions for further research. In an attempt to focus, however, the models suggest a structured approach that helps to pull together strands of the foregoing chapters and to present key issues raised, for which comparative research will better inform our understanding of the relationship between tourism development and wider transformation processes. Such research will, in its turn, redefine and refine the models in an iterative process.

First is the issue of market positioning, image and identity. Several paradoxes have emerged in this volume concerning the relationships between the search for national identity, re-awakening of ethnic particularism, images of authenticity, the domestic need for modernization, and the roles of tourism in facilitating, articulating or compromising such sometimes

The societal context
Post-communist societies' transition sequence
From subsidized domestic and prescribed inbound and outbound international tourism to unsubsidized domestic and unfettered international inbound and outbound tourism

The structure and spatial nature of the tourism industry	
Characteristics of the tourism transition model	*Some challenges of the post-communist transitional response*
Change in the dominant mode of tourism activity	• Decline in domestic tourism as subsidies cease • Economic leakages as outbound tourism market develops • Early rapid expansion of mass tourism confronted with inadequate infrastructure and insufficient expertise • This is followed by a decline in inbound numbers, partly in response to negative experiences and the 'one-off' nature of inquisitive tourism, and then by fluctuating relative stability • Reliance on (not necessarily appropriate) development expertise from 'the West'
Balance of tourism industry organizational structure	• Increasing role for the private sector but within a skills and enterprise vacuum • Changing role for the public sector from provider to enabler, with a resulting loss of power for public officials (although the same individuals may hold additionally new private-sector roles), and resulting protectionist policies • The continued, often hidden, importance of pre-existing networks • Growth of NGOs in the tourism development sector
Scale of tourism industry organizational structure	• The transfer of state concerns to large national and international chains, with the purchasing power but with little or no experience of the tourism industry • The growth of micro-provision, especially in rural areas • Personal and group networks may act to link or polarize structures of different scales
Spatial characteristics of tourism activity	• Increasing diversity of tourism environments often modelled on Western exemplars • Potential for competition and complementarity where large-scale developments are juxtaposed with rural or coastal niche products • Early growth of inbound tourism is unsustainable, tending to polarize around major centres and thus exacerbating unequal regional development prospects

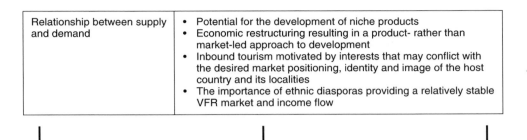

Relationship between supply and demand	• Potential for the development of niche products • Economic restructuring resulting in a product- rather than market-led approach to development • Inbound tourism motivated by interests that may conflict with the desired market positioning, identity and image of the host country and its localities • The importance of ethnic diasporas providing a relatively stable VFR market and income flow

The EU accession context	
Characteristics of EU accession	*Implications for the tourism transition model*
Compliance with the *acquis communautaire*	• Tourism industry structure modified in relation to administrative and legal requirements, including greater transparency of relationships within and between elements • Compliance with EU norms may result in the loss of distinctive identities and the rise of a pan-European blandness • Standardization may also serve to stifle innovation and entrepreneurship, the substance of tourism development
EU territory conceived as a safe European core	• Likely rejuvenation of both mass and niche inbound activity from EU25 markets
Improvements in domestic living standards	• New patterns of domestic and outbound tourism activity to EU15 destinations: more activity, higher levels of spending
Less restricted outbound movement of labour	• Attraction for mobile labour of pre-existing EU tourism centres, possibly acting to hold down wage levels in those centres • Labour shortages may arise from labour out-migration from the country, from less-developed to more-developed regions, and from rural to urban areas, with implications for the regional balance and product mix of tourism attractions and their role in regional development

Fig. 15.1. Tourism's transition within the post-communist transition sequence.

conflicting aspirations. Commodification of the past and present for the purposes of visitor entertainment can create a dissonance that reveals trenchant and rancorous opposition to tourism development and, as a result, to tourism and to tourists themselves.

Related to this is the second point, the relationship between, and issues surrounding, heritage and modernization. Adoption of 'heritage' as both a leisure resource and cultural symbol of identity highlights the irony of employing the (often distant) past as an element of restructuring for the future, particularly for those states wanting to forget the

immediate past decades. This issue has, of course, long been debated in a number of contexts (e.g. Hewison, 1987).

Whilst having broader application, issue number three has a critical relevance for both of the above points. The roles of aid, assistance, consultants and training as catalysts for culturally sensitive, truly sustainable tourism development, continue to require close scrutiny and critical evaluation. Throughout the past 15 years, transfers of know-how and technology (predominantly from the EU and the USA) have 'supported' economic, social, civil and political restructuring processes despite very lit-

tle debate and analysis of the suitability and
sustainability of such processes (see Simpson
and Roberts, 2000). The need to temper consultancy work with academic reflection remains
a critical one for development processes, not
least in relation to the sensitivities of national
cultures and identities.

A fourth set of issues is those concerned
with the consequences of, and strategic
responses to, decision-making processes for policy concerning the economic and demographic
openness of society. The roles of tourism in
conjunction with exposure to world market
forces, the competition of FDI, unfettered flows
of migrants and tourists both into and within
countries, or in relation to the pursuit of protec-
tionism in various forms require closer scrutiny.
A number of chapters have raised issues concerning the porosity both of international and
internal boundaries, particularly in relation to
FDI and to ethnic and social change, and their
roles in tourism development. The value of
comparative (and inter-disciplinary) research is
important. Tourism's role may well be informed
by wider processes and considerations of
investment and development.

Fifth, still relatively little is known about the
roles of ethnic diasporas as important sources
of both tourists and remittances. Such flows
appear to be particularly important for Poland,
south-eastern Europe, northern Cyprus, China,
Vietnam and Cuba. Tourism markets generally
are volatile, being subject to a wide range of
social, political and economic influences. The
global expansion of capitalism, however, not
only provides a transformational context for the
analysis of tourism development but gives rise
to a particular tourism form – the visiting of
friends and relatives (VFR), which has the
potential to provide growing and stable visitor
markets as labour mobility increases.
Something of these markets is known at local
levels but international studies are still required
to provide insight into an increasingly valuable
sector of the global visitor market.

A sixth issue relates to the nature and
importance for tourism development of vertical
as well as horizontal networks and collaboration. In particular we need better to understand
the nature and importance of the 'safety networks' pervading business–bureaucracy collaboration in some post-communist societies.

Seventh, there persists the oft-quoted
(Pigram, 1993; Butler and Hall, 1998a: 254,
1998b: 106; Roberts and Hall, 2001: 71), yet
still poorly documented, 'policy implementation gap' between the nature and context of
legislative frameworks, particularly for sustainable development, and what actually happens
on the ground in the development of tourism.
Tourism policy impact monitoring and feedback processes appear poorly developed in
many tourism destination countries. Indeed,
such an observation relates not only to restructuring, transitional and transforming countries
but to the ostensibly 'model' countries that are
EU members (noting the comments regarding
fitness for membership in Chapter 1). An
exploration of enabling administrative structures in a range of developed and less-devel-
oped economies would therefore prove fruitful
in attempts to narrow the implementation gap.

Eighth, much research is needed concerning
the 'otherness' of race, gender, sexuality, age
and ableism and their roles as determinants of
tourism power dimensions within transformations (Morgan and Pritchard, 1998; Pritchard
and Morgan, 2000; Visser, 2002). The intersection of two or more of these dimensions is of
particular interest: for example, the interaction
of gender and ethnicity has raised critical issues
for tourism employment development. Such
research may have particularly important implications for tourism development in rural areas,
and in contributing to our understanding of the
ways in which issues of core and periphery and
mutually reinforcing components of exclusion
may be recognized and addressed.

Finally, there is the (often discordant) aspiration of all tourism authorities to maximize
income while minimizing the adverse impacts
of tourism. Despite the limited data at our disposal, what is evident, from compilations such
as Table 15.1, is that few destination countries
which fall within the purview of this book are
actually managing to raise their *per capita* level
of international receipts: several reveal an
unstable trend while others exhibit notable
decreases. This fact supports the earlier sugges-
tion of the need to question the nature of consultancy assistance that has been given to many
transforming countries. The continuing challenge to tourism policy makers, across the
range of societies in this volume and beyond, is

to develop a portfolio of sustainable products and services whose sustainability translates into cultural well-being for both host and visitor, environmental protection and enhancement, enhanced and equitably distributed income, and political stability. Or is this a fantasy that must await a yet unforeseen transformation?

Table 15.1. Selected countries in transition: international tourist receipts *per capita*, 1990–2000.

Country	International tourist receipts *per capita* (in US$)						
	1990	1995	1998	1999	2000	2000 1995 as %	2000 1990 as %
CEE Post-communist 2004 EU entrants							
Czech Republic	58	851	679	541	503	59	867
Estonia	na	666	643	589	459	69	na
Hungary[a]	40	135	209	236	220	163	550
Latvia	na	38	319	241	na	na	na
Lithuania	na	118	323	387	318	270	na
Poland	105	344	423	340	351	102	334
Slovakia	85	688	543	470	411	60	484
Slovenia	1109	1485	1110	1084	878	59	79
Mediterranean 2004 EU entrants							
Cyprus	806	851	764	773	705	83	87
Malta	570	589	560	558	535	91	94
Other CEE (including Turkey)							
Albania	133	1625	1929	na	na	na	na
Belarus	na	144	62	na	na	na	na
Bosnia-Hercegovina	na	175	233	233	155	89	na
Bulgaria	201	136	362	377	385	283	192
Croatia	242	905	607	654	473	52	195
Georgia	na	na	1322	1042	na	na	na
Macedonia, FYR	80	127	94	117	168	132	210
Moldova	na	133	200	200	235	177	na
Romania	35	214	88	79	111	52	317
Russia	na	419	412	406	na	na	na
Serbia/Montenegro[a]	352	183	125	112	na	na	na
Turkey	672	700	801	755	796	114	118
Ukraine	na	1039	534	502	na	na	na
Other Former Soviet Union							
Armenia	na	500	333	675	1500	300	na
Azerbaijan	na	777	260	135	na	na	na
Kazakhstan	na	na	na	na	na	na	na
Kyrgyzstan	na	125	136	na	na	na	na
Tajikistan	na	na	na	na	na	na	na
Turkmenistan	na	na	640	na	na	na	na
Uzbekistan[a]	na	na	77	na	na	na	na
USSR	382	–	–	–	–	–	–
Other Asian							
Cambodia	na	454	572	514	489	108	na
China	217	436	503	521	520	119	240
Korea, DPR	na	na	na	na	na	na	na

Continued

Table 15.1. *Continued*

Country	International tourism receipts *per capita* (in US$)					2000 1995 as %	2000 1990 as %
	1990	1995	1998	1999	2000		
Lao PDR	300	417	400	373	380	91	127
Mongolia	33	210	165	176	*na*	*na*	*na*
Vietnam[a]	340	64	57	*na*	*na*	*na*	*na*
Post-apartheid South Africa[a]	963	454	548	419	*na*	*na*	*na*
Latin Caribbean Cuba	736	1320	1130	1099	1033	78	140

na, data not available; [a] arrivals data include all visitors.
Sources: WTO (2002a,b); author's additional calculations.

References

Butler, R.W. and Hall, C.M. (1998a) Conclusion: the sustainability of tourism and recreation in rural areas. In: Butler, R.W., Hall, C.M. and Jenkins, J. (eds) *Tourism and Recreation in Rural Areas*. John Wiley & Sons, Chichester, UK, pp. 249–258.

Butler, R.W. and Hall, C.M. (1998b) Tourism and recreation in rural areas: myth and reality. In: Hall, D. and O'Hanlon, L. (eds) *Rural Tourism Management: Sustainable Options*. The Scottish Agricultural College, Auchincruive, UK, pp. 97–107.

Felker, G.B. (2003) Southeast Asian industrialisation and the changing global production system. *Third World Quarterly* 24(2), 255–282.

Graham, B. and Hart, M. (1999) Cohesion and diversity in the European Union: irreconcilable forces? *Regional Studies* 33(3), 259–268.

Grix, J. and Knowles, V. (2002) The Euroregion and the maximization of social capital: Pro-Europa Viadrina. *Regional and Federal Studies* 12(4), 154–176.

Hewison, R. (1987) *The Heritage Industry: Britain in a Climate of Decline*. Methuen, London.

Morgan, N. and Pritchard, A. (1998) *Tourism, Promotion and Power: Creating Images, Creating Identities*. John Wiley & Sons, Chichester, UK.

Pigram, J. (1993) Planning for tourism in rural areas: bridging the policy implementation gap. In: Pearce, D. and Butler, R. (eds) *Tourism Research: Critiques and Challenges*. Routledge, London, pp. 156–174.

Pritchard, A. and Morgan, N. (2000) Constructing tourism landscapes – gender, sexuality and space. *Tourism Geographies* 2(2), 115–139.

Roberts, L. and Hall, D. (2001) *Rural Tourism and Recreation: Principles to Practice*. CAB International, Wallingford, UK.

Rodriguez Pose, A. and Gill, N. (2003) The global trend towards devolution and its implications. *Environment and Planning C* 21(3), 333–351.

Simpson, F. and Roberts, L. (2000) Help or hindrance? Sustainable approaches to tourism consultancy in Central and Eastern Europe. *Journal of Sustainable Tourism* 8(6), 491–509.

Smith, M. (2000) Negotiating new Europes: the roles of the European Union. *Journal of European Public Policy* 17(5), 806–822.

Visser, G. (2002) Gay tourism in South Africa: issues from the Cape Town experience. *Urban Forum* 13(1), 85–94.

WTO (World Tourism Organization) (2002a) *Compendium of Tourism Statistics 1996–2000*. WTO, Madrid.

WTO (World Tourism Organization) (2002b) *Tourism Highlights 2002*. WTO, Madrid <http://www.world-tourism.org>.

Index

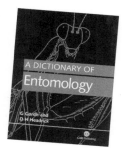
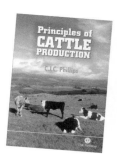